# Five Hundred Horses

# Five Hundred Horses

**Peter Newton**

A.H. & A.W. REED
Wellington Sydney London

*First published 1978*

A.H. & A.W. REED LTD

65-67 Taranaki Street, Wellington
53 Myoora Road, Terry Hills, Sydney 2084
11 Southampton Row, London WC1B 5HA

16-18 Beresford Street, Auckland
Cnr Mowbray and Thackeray Streets,
Waltham, Christchurch 2



ISBN 0 589 01125 1

Typeset by Printset Processes (1973) Ltd.
Printed by Kyodo-Shing Loong Printing Industries
Pte. Ltd, Singapore

*Front endpaper:* All the big stations bred their own horses. These are Clydesdale brood mares at Russell Dalzell's property at Culverden. *R. Dalzell*

*Back endpaper:* Rider Bob Stevenson, late of Outram, on a real twister. Look at the angle of the horse's back legs. *G. Roberts*

*Title page:* "Those old teamsters did a great job." Four four-horse teams ploughing on James Ashworth's property at Leithfield. *Mrs L. Ashworth*

# Contents

*By the same author:*

Wayleggo
High Country Days
High Country Journey
Mesopotamia Station
Straggle Muster
The Boss's Story
Big Country of the North Island
Big Country of the South Island
Ten Thousand Dogs
In the Wake of the Axe
Sheep Thief (a novel)
Walter Peak
Sixty Thousand on the Hoof

# Foreword

A beautiful little bay gelding standing about 14-3, he was by a well-known thoroughbred sire, Ace of Diamonds, out of a sturdy sort of pony mare. Perfectly coupled, he had the strength and constitution of his dam, and with the head and legs of the thoroughbred he showed the spirit and fire one would expect from such a mating. I first saw him in the spring of the year, when he was looking his best, and that one look was enough. That horse had to be mine.

I soon found that he had faults aplenty, but for all that he proved to be by far the best and gamest horse I've ever ridden. I had him for ten years or more and it is mainly my memories of that little horse, Jimmy, that inspired me to write this story.

I've been a shepherd and stockman most of my working years, and horses have been a part of my life for almost as far back as I can remember. I have never reckoned myself a good horseman but I have owned, and known, some wonderful animals. From the immortal Cardigan Bay; the notorious outlaw Aristocrat; the dirtiest of pack mules. Down to Bert Newton's old sledge mare, Jess, I've known them in their hundreds – fast one, slow ones, good ones and bad ones. I feel that I have a story worth the telling. To me the horse ranks next to the dog as man's best friend, and I welcome the opportunity to pay him this tribute.

Peter Newton
Rangiora 1977

# Author's note

AS all these happenings occurred long before we were cursed with metrics I decided to stick to the old system. This, I am sure, will meet with the approval of most of my readers.

However, since many of the younger generation are no doubt as confused by the old way as I am by the new, I include the following metric conversions:

| | | | |
|---|---|---|---|
| 1 in | = 25.4 mm | 1 lb | = 0.45 kg |
| 1 ft | = 305 mm | 1 cwt | = 50.8 kg |
| | | 1 ton | = 1.02 tonnes |
| 1 yd | = 0.91 m | 1 pint | = 0.57 l |
| 1 mile | = 1.61 km | 1 gal | – 4.55 l |
| 1 oz | = 28.3 g | 1 hectare | = 10,000 sq. metres |

I should perhaps also add that although some of the horses and incidents recorded in this book have been mentioned in my earlier writings I felt that they had to be included. Surely the horse is worthy of a book of its own.

Finally, I offer my thanks to all who have assisted me with this work, especially those who so generously made their photographs available.

# Glossary

**Blow his trumpet:** To brag.

**Blocks:** A set of chains and pulleys for yoking a four- or six-horse team to farm implements. They are attached to swingle trees which in turn are hooked onto the implement. They are so designed that every horse in the team does an equal share of the pull.

**Butts:** A small stack of sheaves which has not had the top completed and which consequently will not keep out the weather for any length of time.

**Banjo:** An Australian slang term for a long-handled shovel as used in shingle pits.

**Babbler:** A station cook – sometimes called The Babbling Brook.

**Brumby:** An Australian mongrel-bred wild horse.

**Cocky:** An Australian name for a small farmer.

**Coulter:** Tubes lead from the seed box on a drill down to the coulters which channel it an inch or two into the ground. These coulters are usually 7 in apart and a 17-coulter drill covers about 10 ft.

**Crow:** A stacker's assistant. This is the lightest job in a stacking gang and is usually done by a boy.

**Cob:** A sturdy, comparatively low-set hack.

**Cleanshins:** Cattle which have not been earmarked. The term usually refers to wild cattle.

**Entire:** Stallion.

**Found:** Board and keep.

**Furrow horse:** The off-side (right-hand) leader in a team. So called because when ploughing it walks in the furrow made the previous round.

**Frog:** A tender portion in the centre of the bottom side of a horse's hoof.

**Face:** An open hillside.

**Governess cart:** A horse-drawn passenger conveyance which is entered from the back and has a seat on each side. Made to accommodate four to six passengers.

**Gelding:** A male horse that has been castrated.

**Gorge:** In Canterbury most of the bigger rivers have their source in the ranges. They usually emerge out onto the plains through a narrow gorge and the area beyond this narrow confine is referred to as the gorge; e.g., mustering up the Rakaia or Rangitata Gorge.

**Gut:** A narrow little gully which has probably been washed out by heavy rain.

**Hands:** A horse's height is measured by hands. Each hand is four inches, approximately the width of a man's hand.

**Hacking:** To ride a saddle horse.

**Hames:** A metal brace which is strapped to the collars of team horses. It is equipped with steel hooks onto which the chains can be attached.

**Leading dog:** A dog which stops out in front of the mob of sheep and prevents them from travelling too fast. These dogs would stop there and lead a mob all day and were essential in a drover's team.

**Lawyers:** A bush common in the South Island high country. It is characterized by long tentacles covered with little "hooks". These hooks catch in the wool, and sheep are often unable to pull themselves free. Where lawyers are bad it is necessary to ride around the area looking for sheep which are hung up.

**Lazy stick:** Pulling on the lazy stick is a test of strength. Two men sit on the floor facing each other and with the soles of their feet placed against the other's. They then grasp a stick placed over their feet. The winner is the one who has strength enough to pull his opponent clear of the floor.

**Moisture test:** Grain has to be tested for moisture content before it is suitable for milling.

**Outstation:** Some of the bigger stations establish a base camp in an area far removed from the homestead. This is known as an outstation.

**Packman:** The man who looks after the packhorses and carts the musterers' gear from camp to camp and also cooks their meals.

**Pigskin:** Saddle. To travel per horse is to travel per pigskin.

**Pastern:** The small joint between a horse's hoof and its fetlock.

**Striking out:** Ploughing is usually done in lands (strips two chains wide). When starting a land the first furrow or furrows are known as the strike-out. From then on the ploughman works up and down the paddock turning all furrows in towards the strike-out until the "land" is completed. He then makes another strike out.

**Stook:** Before being stacked sheaves of grain have to be stooked for a week or two. Each stook consists of 12 sheaves, six a side. They are placed in the form of an inverted V, heads uppermost and left until the grain has hardened (or the moisture content is right) before being threshed or stacked.

**Stand:** Each shearer has his stand where he actually shears the sheep. His stand is opposite the porthole which opens into the outside pen where he lets his sheep go when shorn. A shearing shed may be a "four-stand shed". Big sheds run to a dozen or more.

**Walking country:** Country which is too steep and rough for horses. This is the case with most of the South Island high country.

**Yakka:** A slang term for work.

# 1. Schoolday memories

AS a city boy brought up in the heart of Christchurch I had little to do with horses until I left school, but somehow or other I always evinced an interest in all stock. My first association with horses dates back to pre-World War I days when, at the age of six, I used to visit an uncle, Bert Newton, who was farming at Little Akaloa on Banks Peninsula.

Old Bert – he looked old to me, anyway – milked cows, and a daily job was sledging milk down a narrow cutting to a cheese factory in the bay. He had an aged draught mare called Jess which he used for this job, and at times he'd give me the reins when coming back up. Bert's farm was on the sunny side of the bay, pretty steep country, and although Jess would have gone straight home unguided I thought I was pretty hot stuff steering her round some of these hairpin bends. At times Bert would throw me onto her broad back and, clutching the mane like grim death, I'd ride back home. I used to blow my trumpet a bit when I went back to school. I don't suppose any of my cobbers had ever been on a horse, let alone on steep rough country like that!

From then on horses really meant something. I used to admire the milkman's horse, marvelling at the way it would quietly plod to each gateway and stand obediently while each can or billy was filled. Sometimes I sneaked out at night to have a look at the night-man's horse – a draught not unlike old Jess. However, this came to an abrupt stop after an unhappy episode with the night-man himself. Our house, in Brittan Street, was on a little hill. Access to the "little house" was up a steep narrow path at the back of the house. One dark night one of my brothers and I decided we'd play a joke on the night-man and while he was busy getting the pan out of the dunny we tied a rope, ankle high, across the path.

As I said, this resulted in a most unhappy and unpleasant episode – although at the time we thought it hellish funny. Needless to add, we changed our minds the next morning – my old man was pretty good with a switch.

It was about this time that three old maids, named Dorman, occupied the house next door to us. They had a brother, Bill, who was farming up Springfield way – under the slopes of Mt. Torlesse – and at times he came in to visit them. He drove a pony in an old governess cart and used to come in from his place 40-odd miles away and return home the next day. He was a good old bloke and sometimes when he was heading back he'd give me a ride for half a mile or so. If there wasn't much traffic about he'd let me drive. As it happened I got to know Bill Dorman – and his two sons Gordon and Ron – well in later years and we often talked of that pony. She must have been a tough, game little mare.

It was soon after this that the big house opposite our place was purchased by people named Peter from Kaikoura. They brought a grey pony with them – it was a little beauty – and ran it in a section alongside the house. One of the boys rode him and they competed at the country shows with, I believe, quite a bit of success. He looked a champion to me and I spent half my time hanging over the fence admiring him; but that was as far as I got. I'd have given a month's pocket money for a ride on that pony, but as far as that Peter's kid was concerned I might never have existed.

From then on I became a bit mad on horses. At that time, of course, practically all transport was horse-drawn. Even in the cities there were horses galore. Road construction was done with horse and dray – three heavy draughts in the old tip drays – and a local contracting firm, Brightling Bros., had their establishment only half a mile or so from our place. They had a dozen or so drays and some splendid horses.

Haywards, a well-known Christchurch firm of general carriers, had their headquarters right in the heart of the city. Their big pantechnicons – enormous enclosed waggons especially designed for furniture removal – were a familiar sight. Some years later I worked with a chap who had worked for Haywards. He told me that in order to stop drivers loitering while on their rounds it was a set rule that the horses had to be kept at a trot; if anyone was caught walking horses while on the road there'd be trouble. However, some of the drivers, not to be outdone, became experts at schooling their horses to trot even slower than they walked. Although going through the motions of trotting they simply picked up their feet and put them down again in almost exactly the same place – marking time at the trot. This suited the horses as well as the men; none of them were looking for work.

There was no end to the use of horses about the city – bakers' carts and milk carts, each of special design, gigs, buggies and of course saddle horses. Old Harry Box, a well-known Christchurch horseman, who for many years assisted in the selling ring at H. Matson and Co.'s fortnightly horse sale at Tattersalls, drove through Cathedral Square as regular as clockwork. In a beautifully turned out horse and gig, with his bun hat and whip in hand, he was a real showman and could not help but attract attention. In comparatively recent years old Bob French, a well-known drover, once rode his horse up to the doors of the Bank of New Zealand in Cashel Street. Nonchalantly tying it to a lamp-post, he strolled into the building to cash a cheque. The hitching rails and water trough, common at one time, had long since gone.

But most intriguing of all were the old hansom cabs. The cabby perched high in his seat at the back looked out over the box-like cab as he threaded his horse through the traffic – lord of all he surveyed. Personalities in their own right, those old cabbies took a real pride in their calling; the cabs, invariably painted black, were kept in spotless condition and there were some beautiful horses. One old cabby, "Shiny Bob" Allen, became almost an institution in Christchurch. He was many years on the job and if memory serves me right he was the very last of them to ply that trade. To us boys he vied, in importance, with the city mayor. Unfortunately I could never afford a ride in one of those old cabs. Even in those days the threepence a week I got for pocket money didn't run to that.

On Saturday mornings I'd sometimes join up with Terry Jones, one of my companions in crime, and we'd bike to Wildwood Farm down New Brighton way. This well-known trotting establishment was the home of Wildwood Junior, one of the greats of earlier days – or so Terry told me. We used to watch up-and-coming horses flashing round the dirt training track; as they pounded past, nostrils flaring and in a lather of sweat, they all looked champions to me. Terry was only a year or so older than me and where he got his knowledge of racing I don't know. But he seemed to have all the answers.

Yet perhaps best of all were the Sunday outings when we went to Hagley Park to watch polo. The polo ground was straight opposite the Addington Saleyards and was a lovely spot. We thought this a terrific game. After each chukker we'd lead the horses round while they cooled off. Although this meant that we'd miss most of the game it was a good source of revenue. The players used to give us a threepenny bit – and sometimes sixpence – for doing this. Terry always had an eye for business, but I'd sooner have watched the games. Those little horses – they seemed smaller than the present polo "ponies" – were marvellous. To add to the excitement there were some spectacular spills, but those chaps could certainly ride. How they stopped in those little saddles, as shiny as glass, I don't know.

Another highlight was the occasional visit to the horse sales at Tattersalls. If I remember rightly this was held fortnightly on Tuesdays, and we could only go during school holidays. The Tattersalls establishment, with its front entrance in Cashel Street, almost next door to Tattersalls Hotel, ran right through to Hereford Street. Of course it was right in the very centre of the city. There was always a big line-up of horses – anything up to 50, I'd think – mainly hacks and harness horses, although there were the odd special sales of trotters and blood stock.

H. Matson and Co., whose offices were in the front of the building, conducted the sale. Lester Matson, a fine old gentleman, was the big chief; his two sons, Alan and Jack, did most of the selling. Alan was later to become president of the Metropolitan Trotting Club. They were both great auctioneers. Jack, in particular, noted for his quick wit and engaging manner, was reckoned one of the best in the game. It used to be said that if you had something to sell you should give it to Mastons; if you were a buyer it was another matter. There were many stories accredited to Jack Matson, one of which so tickled my fancy that I must tell it. Jack died many years ago so it should be quite safe now.

On this occasion Jack was selling household effects, and among the chattels was a big saucepan. Pointing out the article concerned he started the bidding. "One and three, one and three, thank you Mrs, one and six, one and six, one and six," he chanted. The bidding was then interrupted by one old girl who pointed out that it had a hole in the bottom.

Jack never hesitated. Without missing a beat he carried on.

"So have you, so have I, two in yours and one in mine, one and nine, one and nine, any advance on one and nine."

Those days at Tattersalls were full of interest. Little Harry Box was there for many years, but the real personality was old Jock McGregor. He was tall and spare with a little goatie beard, and Tatts wouldn't have been Tatts without old Jock. With his gigwhip flicking round the horses' hocks and his high-pitched "Ap, ap; Ap, ap", he'd have old has-beens stepping round the selling ring like two-year-olds. Jack lived right alongside our school – Linwood – and he had horses coming and going almost daily. Many years later I worked with his brother Dan and we spent many hours reminiscing.

The Christchurch Show was another outing I never missed and it was again the horses that were the main attraction. In those days – the day of the horses – there was a great line-up of Clydesdales, and there were some beautiful animals. The great stallions, towering a full 17 hands and more, beautifully turned out, were truly magnificent; as though appreciating the importance of the occasion they pranced round the ring full of power and fire and to me were the star attraction of the whole show. The mares, also, were strikingly beautiful; many of them had foals following them, some only a few weeks old. They were the loveliest of all.

And there were the saddle horses, almost in their hundreds – from the big "16-stone" hunters to the tiniest of ponies. The saddle horses headed the grand parade and were an impressive sight.

But to me the Clydesdales – now almost forgotten – took pride of place. Those mighty entires stood out in their magnificence above all else. In later years I served my time behind a six-horse team and came to appreciate the full worth of the working horse.

Another attraction was the side-show run by the famous Australian aboriginal rough rider, Queensland Harry. With a string of outlaw horses he offered good money to anyone who could ride them, and there were many furious minutes while local champions tried their skill. Queensland Harry himself – then reckoned one of the best riders in the world – also gave exhibition rides. His prowess on a horse had me gaping in admiration. He seemed to be a part of the horse itself, and it seemed virtually impossible for any of them to unseat him. He was far from a young man then too – but oh boy, could that man ride!

After the grand parade Terry and I usually sneaked round to the shunting yards at the back of the Addington trotting grounds. Climbing up on top of an empty sheep truck, we got a good view of the last two or three races. There was a bookie operating round there and Terry sometimes had a small bet; however for all his vaunted knowledge he never seemed to make any money. But it was a good day, Show Day, and it never cost us much.

The really great event in those school days was when I acquired a horse of my "own". I got up early one morning to find a stray horse grazing on the roadside. It was unknown to me, so getting a bit of rope from the tool shed and enlisting the help of one of my brothers I tried to catch it. But the horse wasn't having any, so we finally chased it through a gateway into someone's garden. We had it cornered then and although it made a bit of a mess of the garden we finally caught it. Fortunately the owners of the house weren't early risers – and must have been pretty good sleepers too – and we got it away unseen. From what I heard later it's just as well too. This chap, old man Dawber, was a bit fiery, and by all accounts he really went haywire when he saw the damage. It wasn't that flash a garden either.

Anyway we sneaked the horse away to an empty section – there were plenty about at that time – and put it in there. Well, I had that horse for the best part of a week and it was just the job. Terry got a bridle from a boy whose father had a milk round and at night, after school, we'd take it out and ride it up and down the road. We were always joined by several other kids, all wanting a ride. As the "owner" I felt that I was a person of some importance.

However one day, when we were hacking it around, the local cop happened along and feeling that there was something amiss he started making enquiries. I told him that it was my father's horse and that I was supposed to be feeding it along the roadside. Fortunately he took the matter no further but he told me to get it off the road – and keep it off. He reckoned it was a hazard to passing traffic. I didn't argue the point.

I didn't get much sleep that night through worrying – I didn't think they were hanging people for horse stealing at that time but I wasn't sure and I wasn't taking the risk. Anyway I got up early the next morning and letting it out of the section I chased it away to hell down a back road. I was sorry to see it go for we'd had a lot of fun with that horse – Gloaming we called it. It was a nice horse.

Terry went mad when I told him. At the time we were doing a bit of cattle dealing; we'd buy a calf at the Addington saleyards, snig it home on a rope and turn it in an empty section. After a couple of months we turn it over and get another one; we made thirty bob on one and it looked like being a good caper. Terry reckoned that when we got going in a bigger way Gloaming would have been pretty handy for droving. Those calves took some snigging and Terry had too many brains to do hard work.

He said the horse wasn't actually stolen and anyway no one had ever been hanged in this country for horse stealing. But it was alright for him – it wasn't his neck that was at stake.

Above: Six binders operating on the farm of Mr James McIntosh before the turn of the century. The drivers are six of his sons. *F. McIntosh*

*Left:* You could pile just about anything on them. Pack-horses during an autumn muster at Mesopotamia in the early 1930s. *W. Nimmo*

*Above:* "In those days there was a smithy in practically every township." The Littles' old blacksmith's shop at Woodend. Jack is holding the horse's leg, while in the centre stand George and James (hands on hips), the father. *Mrs A. Morriss*

*Below:* You seldom see a sight like this today. Oatstacks on a Rangiora farm. *Rangiora Museum*

# 2. Behind a six-horse team

IT was soon after this that my parents decided that it was a waste of time trying to educate me. That meant that I had to find a job and naturally nothing but farm work would do. It was the general belief in those times that if a young fellow didn't have any brains that was the place for him anyway.

My Uncle Bert had sold his Banks Peninsula place and was then away up in the sticks, somewhere north of Hokianga. So he was out – and I'll bet he wasn't upset about that either. However I'd recently spent a few days with a cocky, Charlie Leach, out Rangiora way, so my father approached him. As it happened he wanted a cowboy and against his better judgment he gave me the job – $2 a week, sleep in the whare with the men, tucker with the married couple. Not very flash tucker either – but it was a job on a farm.

I soon found that milking cows and looking after the vegetable garden wasn't much of a catch. But I must have shaped reasonably well – or else the boss was too soft to sack me – because when a few months later one of the teamsters rolled his swag and took off I got that job. It was only the third team – a pretty scratch lot – but I was a teamster. A vast improvement, socially anyway, on milking cows.

It was then that I really started to learn something about horses. There were three six-horse teams on that place and with just on 600 acres in crop, as well as sheep feed, they were kept going all the year round. The two top teams – No. 1 and No. 2 – were crackers. I can't remember much about the individual horses – it was over 50 years ago – but the furrow horse was a big bay gelding called Major and he was a beauty; as brainy as a man. Another that I remember was a chestnut mare, by Black Douglas, and a tiger to go. She worked in the middle, in the lead, and although she was always tied back her nose would be a foot in front all day. She was a mad-headed sod and could be a bit of a problem at times, but talk about guts.

My own team, as I said, were a pretty scratch lot. Mainly older horses and the odd one that didn't measure up. One in particular, a mare called Stilts, was adflicted with "string-halt"; about the worst cast I've seen, but for all her disability she was a great toiler. This team did only the "top" work – harrowing, rolling, discing, and so on – and although they didn't compare with the other teams they did a lot of work.

All told there were 20-odd draughts plus three hacks on that place. Most farmers bred their own in those times and as well as the working horses there were several young-ens – yearlings and two-year-olds. I remember the hacks well. One was a bright bay named Dolly; comparatively young and a bit lively, she was a lovely mare. Another, Ginger, was a chestnut, getting on in years but a nice sensible hack. The third one, a quite remarkable horse, was said to be by a son of the immortal Carbine. I can't remember his name. Although obviously a clean-bred horse, he had a great top – a typical 16-stone hunter. Just a ball of muscle, he was as strong as a lion. He had been broken into the saddle and, apparently, was an outstanding hack; but soon after that a plough share, hidden in long grass, had pierced the frog of his foot and ruined him for road work. But the boss refused to "dog tucker" him and he was just kept as a spare team horse. He also worked in the middle in the lead and although he looked a midget alongside the big draughts you just couldn't kill him. Like the chestnut mare, he had his nose in front all day and fairly put the big horses to shame.

Dolly was used in the spring cart mainly, on fencing work and odd jobs, and I drove her quite a bit. On one occasion when I was coming home with an empty cart I spotted one of the old 4 gallon kerosene tins lying out in the paddock. Trying to be smart, I ran over it with one wheel. I hit it dead in the middle and as I might have known the tin buckled round the rim of the wheel. There was one hell of a clatter and Dolly, always a bit touchy, promptly bolted. The clatter continued every time the wheel went round and with the empty dray bouncing around like a yo-yo and the rattling of the fencing tools I never had Buckley's show of holding her. We did about half a dozen rounds of the paddock before I got her stopped. The problem then was to get that bloody tin off the wheel without startling Dolly again. In the end, I had to unyoke her and tie her to the fence.

Fortunately the boss was away at the time, so there were no repercussions, and if nothing else I'd had a pretty thrilling ride. I reckon that for the first couple of rounds that mare had been touching

50, and at times the dray was bounding three feet in the air. Dolly would have won a trotting cup that day.

As I remarked earlier, most farmers bred their own horses in those days. Entires led by a man on horseback travelled the roads from September to Christmas and the mares were usually served on the farm. Some that I remember, of about that time, were Baron Bold, Black Douglas, Streamline and Baron Belmont. This was then big business, several horses often doing the same district. Some of the grooms followed the game for years and became widely known. Posters, complete with a photograph or drawing of the horse and explaining its breeding and virtues, were sent to most farmers. They were usually tacked up on the stable walls. In some stables the walls were almost covered with them.

However on one occasion the groom didn't call in at Leach's, and I had to take a mare along to a neighbouring farm. She was a flighty sort of a thing and when the stallion was introduced she really went to market. The groom had no option but to rope her in case her flying hooves did some damage, and while he was doing this he got me to hold the stallion. At the time I was only about as high as a horse's knee and that stallion, all of 20 hands – it seemed to me – was pretty fearsome. With the scent of things to come he was rearing and snorting and plunging like a mad thing. He spent most of the time walking on his hind legs and, with his front feet flailing the air about 20 ft above me, I was sure I'd be battered to death any minute. The groom, who was having his own troubles, was screaming at me to hang on to him – about 2000 lb of mad horseflesh – and I can tell you I wasn't too happy.

But the mare was finally roped up and we got the job done. The funny part was that all this was a complete waste of time as the mare didn't hold.

Another job I had, during the harvest, was helping with the horses used in the binders. These horses worked a two and half-hour shift – four changes in a day – and my job was to take out the fresh team and then take the others back to the stable and feed and water them. With short shifts like that the teamsters were able to keep them stepping out freely all day. There were two binders operating – three horses to each – and with some of the paddocks well over a mile from the homestead I was kept busy. However I enjoyed that job – anything was better than weeding bloody carrots and onions.

With 600 acres in crop the harvesting was a big job. In those days most of the grain was threshed out of the stook, although it was necessary to build a few "butts" a week or two before the mill arrived. Stook threshing could not start until all traces of dew had disappeared, and with these "butts" to start on in the morning the mill could get going in good time. With half a dozen cockies all screaming for the mill at the same time every hour counted. The oats, which were mainly for chaff, were always stacked. All told the stooking, stacking and the carting to the mill involved a lot of hard yakka, and a lot of men.

With two binders operating, four stookers were necessary – six men all told. Seven if you counted me. The stacking gang required five: two dray men, one forker in the paddock, a stacker and a "crow". When threshing out of the stook the farmer had to provide four drays and a man to each dray.

This, of course, was well before the day of the header harvester. No moisture tests were necessary as with the grain all in bags it was stored in a shed on the property until it was "ready". The cartage, to the grain shed and then to the railway station, was all done by horse and dray and this itself was quite an undertaking. When carting to the station we used three drays, each with three horses, and the regulation load was thirty 200 lb sacks – the old three stripers. There were somewhere around 5000 sacks all told, and the carting to the station alone took a good two weeks. Although only a midget, I drove one of the drays and used to just about bust my foo-foo valve upending those wheat sacks. However according to Charlie Leach, the boss, I was a strong little fellow – as strong as a lavatory with the lid off.

Starting with the reaping of the oats right through to the carting of the grain, the harvest on that place lasted well over two months. Summer and winter there was little respite for either man or horse.

This was all good experience too. By the time I was given the teamster's job I thought there was nothing I didn't know. However I soon found that driving a team wasn't all it was cracked up to be. The teamster's job starts at 5.30 in the morning when he gets the horses in and feeds them; from then until breakfast, at seven, he has to groom and harness his team, get their mid-day feed bags ready, fill up the chaff bin and if time permits clean the stable. The rule was eight hours "tight chains" (the horses had to be actually working for eight hours) and with some of the paddocks a mile or so away it was necessary to be away from the stable by 7.30. As well as the trip to the paddock the implements had to be oiled and greased and of course the horses yoked up – all before eight o'clock.

Unless working handy to the stable the horses had their mid-day feed on the job. They were simply unyoked, given a drink from a handy water race or trough (their only drink of the day) and fed from "nose bags". The bags, which held a kerosene tin of chaff, were attached to the horses heads by a short strap or length of rope; the one tin of chaff usually kept a horse chewing for the best part of the dinner hour. The teamsters ate their own lunch alongside their horses.

Traditionally this was simply a great bread and mutton sandwich – about 2 in thick – and a bottle of cold tea. With a bit of luck there might also be a slab of cake. With chains "tight" again by one o'clock they worked until five, and it was then the best part of 5.30 before they were back in the stable again. By the time the horses were fed and unharnessed and the men had washed off their layer of dust it was tea time.

However the teamster's day was by no means over. The horses were fed three times – three tins of chaff – before finally being turned out for the night at about 8.30 to nine. During the winter months they were all covered. A 15-hour day, six days a week. Even on Sundays he had to be on the job, the horses being fed morning and night. It was also a regular practice to give the stables a thorough clean out on the Sunday morning. A six-horse team eats roughly a ton of chaff a week – and a ton of chaff makes a hell of a lot of dung. This was usually barrowed out to a heap a chain or so from the stable and was quite a job. Not many teamsters found time to go to church.

All told those old teamsters did a great job, and they did it for little more than their tucker. In my time top wages were $4 a week. I did it for $2.50. Yet some of those blokes spent the whole of their working life behind a team.

Those teams at Charlie Leach's put up some wonderful performances. On one occasion four horses from the No. 1 team drilled 32 acres of turnips in a day with a 17-coulter drill. I don't know just how many hours they put in – it was well over eight – but even so horses are not machines and it was a great effort. The team at that time was driven by a chap named Arthur Greening. On another occasion Bill Kinley, who was driving the No. 2 team, drilled 27 acres of oats in the eight hours. I was carting the seed, which was split into half-sacks for ease of handling, and Bill kept the horses going non-stop. He'd have two half-sacks of oats sitting on the drill and would simply fill up while the horses were going. The horses, of course, followed the mark of the last round unerringly, and while filling up Bill just hooked the reins up and let them go. In order to cut costs – this was during the '21 slump – no manure was used and this was quite a saving in time. I might add that this was in a 70-acre paddock, about half a mile long and with a minimum of turning; but all the same that was another big day. It was usually reckoned that an acre a coulter was fair enough.

I can't remember what acreages were cut with the binders when harvesting, but they'd have taken some beating too. The boss was a great believer in feeding and his horses were always in great fettle. A liberal ration of crushed oats was always added to each tinful of chaff, and this was also dampened with diluted molasses. The horses just loved this and it put a great shine on their coats. Another practice that I've not seen elsewhere was that often in the heat of the summer the horses were given a two-hour spell at mid-day and then worked on till six. The teamsters seemed quite happy with this arrangement although it added still further to their long day.

My job, as third teamster, involved only those jobs that needed little experience (or brains) – rolling, harrowing and grubbing – and most of this required only three or four horses. These worked abreast and were comparatively easy to handle. However I did a fair bit of grubbing and this needed all six horses, working in the "blocks". This was a bit more complicated. A six-horse team doing heavy work, like ploughing or grubbing, works three abreast – three in the lead and three immediately behind. They are yoked in a set of "blocks", a system of chains and pulleys attached to swingle trees and so designed that each horse, no matter how lazy it may be, gets exactly its share of the work. The most valuable horse in the team is the right-hand leader – known as the "furrow" horse. When ploughing, both it and the one immediately behind it walk in the furrow (turned over the previous round). They will do this unerringly without any guidance at all. Even when doing top work the furrow horse can be relied on to follow the mark left by the implement on the previous round without fail. A good one is a real asset.

The most lively horse in the team is usually yoked in the centre of the three leaders and it usually sets the pace. If too free – like the chestnut mare I mentioned – they have to be tied back. Young horses in the process of being broken in are put in the centre at the back where they are firmly held in position by the others. When the horses are unyoked at night the chains are simply dropped on the ground in position. It was uncanny the way each horse would step into its place when being yoked up the next morning.

The main thing when driving a six-horse team in the blocks was to keep the chains tight when turning. If they got too slack the horses in the back were liable to get their hind legs tangled in the chains, and if they were a bit flighty there could be trouble – especially with a young one. I soon got used to this and was a bit inclined to blow my chest out. Needless to add I soon got that knocked out of me.

On one occasion I was harrowing with the six horses abreast. This is a difficult team to handle at the best of times but the six horses abreast are the width of the six leaves of harrows, and it was not uncommon. However the boss had rigged up a little harrow "cart" – an old plough seat set on two small wheels and attached to the harrow bar by a long pole. Although it was very nice to be able to ride, it was hard to control the horses while perched on this seat. Anyway during the day something upset the horses and they broke into a trot. I never had a hope of holding them, and quite expected that as the excitement mounted they'd finish up by bolting. There was a wire fence straight ahead, but down one side of the paddock there was a big gorse hedge. I had to get them turned from that wire fence somehow – there'd be hell to pay if they hit it – and more by good luck than good management I managed to slew them round and head straight for the gorse. That stopped them, but there was a fair tangle for a while. I had to unyoke the lot and drag the harrows, chains and bar out by hand before I could get going again. I wasn't too happy either and decided to have half an hour's spell before I yoked up again. I took the harrow cart off and walked for the rest of the day.

But worse was to come. On this particular occasion I was working about a mile from the stable. When I knocked off at night I decided that instead of leading the team home still coupled together, as is the usual practice, I'd let them loose and ride one of them. Well, I let the others go and they set off down the road nice and quietly. By the time I got my mare alongside a gate and clambered aboard they were some distance ahead. My "hack" didn't like this and set off after them at a good canter. Hearing her pounding along behind them, the others took off flat out with me hard in behind them. I had no show of holding my mare and we landed in the yard full gallop with the team in a lather of sweat.

As it happened the boss was about that day – right at the stable door – and this time he made no bones about telling me what a useless, bone-headed, lazy and irresponsible so-and-so I was. I didn't know I had so many faults and at first I thought he was just joking. But no fear, he meant every word of it. Anyway on thinking it over I decided that it might be a good idea if I had a change of grazing. The upshot was that I decided to hand in my portfolio.

My parents weren't too pleased when I landed home minus a job, but I told them that I just couldn't stand the boss's bad language. My mother agreed that if he was like that I was better away from the place. However I felt that the old man was a bit sceptical.

# 3. Horse sales and ploughing matches

IT was while at home – resting from my labours – that I first went to the monthly horse sale at Rangiora. This was an important fixture and there was always a good line-up of horses. Unlike Tatts in Christchurch, they were mainly draughts and some of them brought big money. Up to $160 was not uncommon – a lot of money in those days. St James Station of Hanmer Springs sent down a draft of unbroken horses every autumn and they were always a feature of the sale. In those days St James was breeding a light draught and they were in great demand. Clean-legged horses with a great top, they made good team horses. Although lighter and smaller than the average draught, they put the bigger horses to shame when it came to hard work.

They were driven from St James to Culverden, some 30 miles, railed to Rangiora and then driven through the town to the saleyards. Quite a performance at times. Whether they were ever handled before leaving St James I don't know, but there were some fair hoolies when it came to taking single horses away after the sale.

Dealers played a big part in those early horse sales and it was commonplace to see a mounted man leading five or six horses from one sale to another. Linc Wright was one of the best-known dealers in this part of Canterbury and he always had a string of horses at Rangiora. The Binnie boys of Sefton worked for him for several years, collecting horses he had purchased privately and also leading them from one sale to another. They had a bobtailed heeling dog called Bob that was a cracker on that job. Snigging five or six horses all day is quite a job, but with Bob snapping at its heels even the worst of leaders stepped along like a champion. If any horse lagged Bob only needed the word and he'd have it by the heels in a flash.

It was about this time that I first went to a ploughing match. It was a fascinating experience. Frank (Mick) Merrin, a neighbour of Charlie Leach, was then a leading Canterbury ploughman and as I sometimes went across to have a yarn with his teamster – young Dick

Burgess – I came to know something of the requirements of match ploughing. Mick Merrin, of course, had his own match plough, and to me it was almost as complicated as his old Hup car.

Match ploughing was then at about its peak in Canterbury and contests were held annually in most districts throughout the province. Joe Wyllie, a member of a family well known throughout the island, both for their ploughing and their Clydesdale horses, tells me that at the Rakaia ploughing match in 1922 he saw 66 teams operating at once, in the same paddock. This was thought to be a record for a New Zealand ploughing match. Joe, who is now living retired in Rangiora, first went to a ploughing match in 1912; he first competed in 1920 and he has been associated with the game ever since. In 1975 he was a member of the panel of judges who officiated at Rakaia.

In the days of the horses it was customary for four and six-horse teams to plough a quarter and a half-acre plot respectively – the former using a double and the latter a three-furrow plough. There was a time limit of four hours for each section. As well as the various prizes for the actual ploughing, trophies were awarded to the teams themselves – the best team, the best-kept harness and the best decked-out team. Some of them were beautifully turned out, the elaborate harness and fillings being specially reserved for the occasion. They were an impressive sight. What was quite remarkable was the understanding between the teamsters and their horses, particularly when "striking out"; the horses responded to every word and it was an education to see them in action.

Except for odd exhibitions, horse teams at ploughing matches are now a thing of the past. However odd enthusiasts throughout the country are establishing Clydesdale studs and these grand horses are again becoming one of the attractions at many of our leading agricultural shows. It's good to see them coming into their own again.

Although North Canterbury was a real stronghold of the Clydesdale probably no family has played a greater part in the development of the breed than the Wyllies. Joe's father, J. D. (James) Wyllie, came out to this country from Australia with his parents in the seventies; they settled in Otago and it was not until 1901 that James came to North Canterbury. However he immediately made his mark with the horses and for the next 30 years – until his death in 1933 – his record both in the show ring and as a breeder of Clydesdales is probably unequalled. He was also in great demand as a judge and his reputation in this respect was such that he was selected to judge the

Clydesdales at the first Royal Show to be held in this country. This was at Palmerston North in 1923.

In 1906 Wright Stevenson and Co. imported four top Scottish Clydesdale colts and offered them at auction in Dunedin; James Wyllie purchased one of them for 1250 guineas – a tremendous price at that time. It was the opinion of many competent judges that this horse, Baron Everest, did more to improve the standard of the Clydesdales in Canterbury than any other sire we have known. Naturally he was in tremendous demand and one season he covered 120 mares. James Wyllie kept Baron Everest for only four years, but during that time he was the grand champion Clydesdale sire at the Christchurch Show in each of those years.

On James's death his farm and the horses were run by one of his sons, J. G. (Joe) Wyllie, and although the horse teams were then on the way out he continued breeding and showing for several years. Two of his brothers, Jim and John, were also keenly interested in the horses and match ploughing, although Jim was one of the many who lost his life in World War I.

Strangely enough old James Wyllie, Joe's father, married a Wyllie. She was a daughter of John Wyllie of Sefton, who with his three sons, Thomas, John junior and James, were also prominent breeders of Clydesdales. As a family the Wyllies' association with both the horses and the ploughing must be unique.

Going to horse sales and ploughing matches wasn't a bad way of filling in time, but it couldn't last and I soon had to think about getting a job again. I finally got one on a little grazing run, Woodstock, in the Oxford district, but as I had nothing much to recommend me it was a case of taking on the cows again. However this time I had to fill in the day grubbing gorse – worse by far than weeding onions.

Some months after I went to Woodstock a new teamster arrived. Who should it be but my old school-mate, Terry Jones, now – if his talk meant anything – a man of vast experience. Terry must then have been close on 20, and, of course, was then at his most knowledgeable. I hadn't quite reached that stage, but it was soon to come. Terry only stopped a few months at Woodstock before, feeling that he was worthy of better things, he moved on.

The upshot was that once again I was promoted to teamster. What's more it was the No. 1 team – the only one, to be truthful – but even so I doubt if it was any better than Charlie Leach's third team. Like some of those Haywards horses, they could go through the motions without any noticeable progress. The boss pulled up one day and, seemingly quite concerned, asked if I had plenty of grease. He was afraid that the bearings might get over-heated and set fire to the roller! I definitely got the impression I'd have to do something about it; cursing and swearing at the horses had no effect and it looked as though I'd have to do something drastic if I was going to brighten them up. As it happened the opportunity soon came.

On this particular day one of the other chaps, Cliff Cherry, was doing some fencing alongside where I was working. At mid-day I unyoked the horses and had my lunch with him. We'd boiled the billy on an open fire and as it was a bitterly cold day I picked up a stone from alongside the fireplace and was jiggling it from hand to hand in an effort to warm them. This gave me a bright idea. The real slug of the team, a mare named Pryde, worked in the middle at the back of the team. After I had yoked them up and was ready to go I gave Cliff the word and he grabbed a particularly hot stone and slipped it under Pryde's tail. She immediately clamped her tail down and held the stone fast. The next moment she took off like a mad thing. She was well held by the other horses, but even so I reckon that for the first hundred yards or so her nose was almost level with the leaders. For the next few rounds that team would have compared with any in the country.

As far as I was concerned the worst feature of that team was that four of the six were white. At one time the station had kept a white entire and a number of his progeny were white. This horse, Billy, had finally been gelded and was now one of the team. The other three whites, all of them mares, were by him. White horses take the eye; when rolling or something like that I sometimes used just the four whites and they looked very attractive. However in the winter, when there was a bit of mud and slush about, they took too much grooming. I cursed the bloody things. During the winter months the teamster had to groom and harness his team before daybreak and with his only light a single stable lantern – about one candle power – it's very easy to miss the odd sweat mark or mud patch. At times I had to get the brush and give them the once over again after I had coupled them up out in the yard.

I was doing this once when, unobserved by me, the boss arrived in the yard. His remarks were even more sarcastic than at the time of the grease episode. It seemed I just couldn't win.

The six horses in the Woodstock team were the four whites – Billy

Jess, Patch and Snow – the dead and alive Pryde, and a little St James mare named May. A dark brown, clean-legged mare, typical of the St James breed, May was a cracker. In the blocks she worked in the middle in the lead, and no matter how long the day or how hard the going her nose was out in front from start to finish. Although she was by far the smallest horse in the team, the rest of them couldn't hold a candle to her. It was a wonder she didn't go mental having to work with those other no-hopers.

She had originally been purchased as a spring carter, but although normally a beautifully level-headed horse she just wouldn't tolerate working in the shafts. She had smashed the cart up more than once and had finally been put in the team. There was also a bit of packing to do at Woodstock and she was ideal for that job – strong, active, and at only about 15 hands nice and low to load. That's a big advantage unless there is river work. Lifting heavy loads onto a tall horse is not so good. I don't know how it came about that May wouldn't work in harness, as in every other respect she was a lovely mare to handle.

It was at Woodstock that I had my first introduction to pack-horse work. Although then only some 8000 acres, the country ran up to about 4000 ft and for a place of that size the "back country" – if such it could be called – was fairly rough and broken. The mustering involved two days for four men; they camped at "Fosters", about eight miles back from the homestead. As access to this camp was by pack-track only, their bed rolls, tucker and various odds and ends had to go out by pack-horse. This job fell to me, and as the packman also has to cook for the musterers I can't say I was filled with joy. Judging by their comments the musterers weren't either. Cooking was never my long-suit. However we all survived.

This job required two pack-horses – May and a bay gelding named Labour Day. May was a cracker, but of the many pack-horses I've known Labour Day was definitely the worst – a lazy, ignorant, bumble-footed so-and-so. In fact the only time I ever saw him get a move on was when he once lost his footing on the hill and did about ten somersaults down the hill. He was moving on that occasion.

At the time we were salvaging fencing material out at the far end of the run. Woodstock, a long narrow block, lies between the Waimakariri River and the Harewood Forest. Its western boundary is a 14-mile stretch of gorge, running from the homestead to the main bush. To the north and east it is bounded by the bush itself. At one time it was fenced right round the bushline, but out at the far end of the block – in behind Bald Hill – the fenceline ran up to close on 4000 ft. Over the years this stretch had deteriorated beyond repair and it was this that we had to salvage. It was difficult going for horses, but anything we couldn't get them to had to be carried out on our backs so needless to add we took the horses into some pretty dickie places. It was while coming out of one awkward spot that Labour Day came to grief.

Luckily no bones were broken, although he had a couple of nasty flesh wounds, and we were able to get him out and back to camp. A packman's equipment includes a sack needle and twine and my mate Darval Poulton – who had served his time as a bag sower on a threshing mill – stitched the old bloke up like a professional. But that spill finished Labour Day for hill work.

Another experience we had while on that job was packing a live pig. The dogs got onto a pig not far from the camp one evening; it was a boar, about half grown, and Darval got the bright idea of taking it into the homestead to mate with a sow they had. I was all for the idea, providing Darval did the catching. Anyway we finally smothered it with a tent fly, lugged it to our camp and fashioning a wire ring, just above the hock, tethered it to a tree with a dog chain.

That proved to be the easiest part of the job. Getting it onto a horse was a different matter altogether. We decided that May would have to do this job, but she had other ideas – and after her performance that day I could quite understand how she had smashed a couple of spring carts. We decided that we'd first of all load her with our swags and then make a top load of the pig. In theory quite a good idea. After quite a bit of fun we got the pig into a sack, with only its head showing and then came the job of getting it onto the horse. With the pig squealing like all hell let loose May went beserk. But we roped her up until she could hardly move and when she had finally worn herself out we got our pig on board and firmly lashed it into place.

But when we unroped May we found that she was far from done. For a few minutes that pig must have thought all its birthdays had come at once. Fortunately the straps, the circingle and the various bits of rope we'd used all stood the strain and we finally got our pig into the homestead. What's more it duly did its job. The worst feature of that episode was that we'd loaded the pig with its head over my swag and during that 14-mile ride it seemed that it had vomited most of the time. I was reminded of that pig every night for weeks after that.

Woodstock was then owned by R. O. Dixon, who also farmed

several other properties between it and his homestead farm, Holton, situated a few miles west of Kaiapoi. Altogether he did a tremendous amount of cultivation work and his various teams often shifted from one place to another. I once put in several weeks with my team at Claxby, a Dixon property bordering the Waimak some 30 miles downstream from Woodstock. The Claxby team was driven by Harry England, who later became a top Canterbury blade shearer.

We were working very light silty ground out near the river and if ever I experienced dust it was on that job. As usual I landed the two dustiest jobs of all – harrowing and rolling; for most of the time it was practically impossible to see the mark from the previous round and I had no option but to leave it to the horses. My furrow horse, Jess, was pretty good and seldom went wrong. As we were working some miles from the homestead we camped on the job, in a portable whare, and did for ourselves. If ever I earned my tucker it was on that job. Not that I needed much tucker – I had more than my fill of dust by the time the day was done. You can swallow a fair bit of dust in eight hours. But it was good soil and the taste wasn't that bad!

Harry, who was a Pom, was one of the most conscientious blokes I ever worked with – and that's saying a lot when you're talking of those old teamsters. The horses had to be in and feeding at 5.30 in the morning, and they weren't turned out until nine at night. The paddock we were working – some 250 acres – was just a mile long and I was able to do only nine rounds (up and back) a day – five in the morning and four in the afternoon. A total of 18 miles. What's more I had no harrow cart on that job and in soft going, behind harrows, 18 miles is a fair sort of a day. I thought so anyway, even if the boss didn't.

But Harry, who was several years older than me, kept me hard at it. On one occasion, with a six-leaf set of harrows, I did just on 40 acres in the eight hours. That was reckoned to be a record for the Woodstock team. I'll bet old Billy and Pryde skited about that for the rest of their lives. All the same I wouldn't mind a dollar for every clod I threw at them that day. Pryde had long since forgotten the episode with the hot stone and was just as lazy as ever.

It was at Claxby that I first met Jock McArthur, who was also working for the Dixons at that time. I knew him well by name as he was one of the top New Zealand rough riders in his day. His father, John, had once held St James station. R. O. Dixon had once tried to purchase St James himself and he never tired of talking of that vast holding. The second team at Claxby was driven by Frank Gebbie, a member of one of the very earliest of the Banks Peninsula families, and he also was a crack rider. He and Jock delighted in trying out any unruly horses.

It so happened that "R. O.'s" special pride and joy at that time was three beautiful mares that he ran at Holton, his homestead property; for some reason these were never used and had become virtually unmanagable. One of them, a beautiful roan, was just as pretty a mare as any I have ever seen. R.O. himself did occasionally ride one of them, but he was then an old man. That mare, as dirty as they came, was too much for him and he had some almighty spills. But he was as game as Ned Kelly and he usually won in the end.

Jock and Frank would dearly have loved to put her in her place but no one was ever allowed to touch any of those mares. As far as I know the three of them finished their days eating their heads off at Holton.

With the Claxby job finished I returned, with my team, to Woodstock. It was not long after this that I rolled my swag and moved on to fresh pastures. This ended my days as a teamster. My only possible claim to fame was that I had driven what was probably the worst six-horse team in New Zealand and that I had also carted a live wild boar on a pack-horse.

But that interlude behind a team was a great experience and it has left me with the greatest respect and regard for those old teamsters. In earlier days, before the turn of the century, they did an almost unbelievable job. Some of the giant holdings – such as Longbeach and Springfield in the Ashburton county – ran 20 teams and more, and had anything up to 8000 acres under the plough. When the Longbeach partnership of Grigg and Russell was dissolved in the early '80s, over 100 draughts were offered at the clearing sale. The list of agricultural plant was almost beyond belief. With most of the cultivation work being done from plough camps, winter and summer, the teamsters were working under the worst of conditions. Their regulation day was from 5.30 in the morning, when they got their horses in, until 8.30 at night when they finally turned them out. They worked a six-day week and were also virtually tied to their horses on the Sunday.

Yet for all the severity of the conditions under which they worked their love of their horses was such that many of them spent a lifetime on the job. To those worth their salt the care of their horses came before personal considerations and there will be few among the ranks of our workers who are deserving of higher praise. And the team horses themselves are no less worthy. Their whole life was simply long

*Above:* Mr H. E. Cook's six-horse team at a Rakaia ploughing match in the 1920s. *J. G. Wyllie*

*Below:* Seven four-horse teams ready for work. The teamsters are actually the seven sons of James McIntosh. *F. McIntosh* (Ken)

*Above:* The Iron Horse. These engines often took over from the big ploughing teams. *F. McIntosh*

*Below:* An old wooden combine in action. *J. McGiffert*

The magnificent Baron Everest, imported from Scotland in 1906 and four times champion entire at the Christchurch Show. This horse did much to improve the standard of North Canterbury Clydesdales.
*J. G. Wyllie*

Old Cobb and Co. days. Coaches on the Springfield-Arthur's Pass road. *Canterbury Museum*

years of toil – plodding from one end of a paddock to the other, without end. Their only reward was the care and attention they received from their drivers. There were many wonderful animals among them and they also should not be forgotten.

# 4. My first hack

MY next job was in a mustering gang at a place called Winterslow at Alford Forest – known far and wide as The Bush. As a musterer's usual mode of travel was per "pigskin" I felt that I had to get myself a hack. As it happened Darval Poulton had a nice little chestnut gelding which he was willing to sell and we duly made a deal. I'd had to buy dogs also and what with these and Torchy, the hack, I said goodbye to the whole of my savings.

My father, who was then County Clerk for the Eyre County, was stationed at Ohoka, a few miles from Kaiapoi. I had a 70-mile ride through to Winterslow. My main memory of that ride was meeting two drovers with a mob of Merino wethers. They had come from Meadowbank Station, a few miles beyond Blenheim, and were heading for Stronschrubie – now known as Erewhon – at the head of the Rangitata Gorge, a drive of some 300 miles. They told me that they had already been five weeks on the road and with another 100 miles still to go it would be another fortnight before they arrived at their destination. Men, horses, sheep and dogs, particularly the dogs, were all showing the effects of the long drive.

One of the drovers was Walter Grieve, the oldest of seven brothers, all of whom made a name for themselves in the Canterbury high country. Their plant was typical of a drover's outfit in those days – a horse and gig, the latter with a big "basket" slung low beneath the seat and extending a foot or so to the rear. Often simply made of chaff sacks, this was used mainly for carrying the odd sheep that knocked up during the day and for the occasional dog that got too foot-sore to travel. The gig was also loaded with the drovers' swags and the assortment of gear necessary for such a trip. Walter Grieves's mate, Bert Nosworthy, had his hack with him, although when I met them he was perched up in the gig with his horse following behind. When they first started out on the drive Bert would have ridden ahead of the mob as those Merino wethers, fresh and full of gallop, would have pushed a leading dog clean off the road. When I met them – five weeks later – it wasn't even necessary to have a dog in front of them.

At that time – well before the advent of road transport – it was commonplace to see big mobs on the road. Some of the drovers did little else throughout the year. Bert Nosworthy was many years on that job; one of the best-known Canterbury drovers of his time, he followed the game until old age finally forced him into retirement. They invariably travelled by horse and gig and some of their horses became as knowing as a man. It was generally accepted that if the mob exceeded 600 a second drover had to be employed, but it was not uncommon to see one man with anything up to 1000. With a mob that size, on a narrow, winding road, the lead was often out of sight and at times the drover had to leave his horse unattended, while he poked his way up through the mob, on foot, to attend to some hold-up. His horse could be left for long enough, the drover quite confident that it would simply trail along behind the sheep grazing as it went. A good horse was a real asset to those chaps.

Along the main stock routes the various counties had paddocks available for holding stock overnight, and the drovers usually put up at the nearest pub. If no paddock was available obliging farmers allowed sheep to be held in their yards and there was always the chance that he'd have some place where the drovers could camp for the night. The owner of the sheep met any expenses that were incurred and it was also the practice to pay the drovers for any time involved in picking up stock and returning to their homes after the drive was finished. On this particular drive Walter and Bert had to drive to Blenheim to pick up their mob and, when they finished the job, drive from Schrubie back to Christchurch where they were then living. With over 300 miles to travel this alone would have involved just on a fortnight. The round trip, all told, was near enough to 700 miles – and they never even got their names in the paper. Today if some glory hunter rode or drove that distance – just for pleasure – it would be headline news.

The longest droving job that I know of in this country was from Titirangi Station, right on the open seaboard between Queen Charlotte and Pelorus Sounds – some 50 miles from Havelock – to Pigeon Bay on Banks Peninsula. They travelled via Havelock and out to the Murchison highway which they followed to within a few miles of Top House. Leaving the road they turned up the Wairau River, through the Rainbow country, and Hell's Gate, to come out at Tarndale a few miles above the old homestead. They then crossed the Island Saddle and dropping down to Lake Tennyson followed the Clarence River and out over Jacks Pass to Hanmer Springs. From there to Little River, and over the Hill Top to their destination, they simply followed the main roads.

But all told that was a drive of 400 miles or more, and much of it through wild unfenced country. Old Joe Hall, a well-known Marlborough identity who died only a year or so ago, did that drive on several occasions.

Perhaps the most famous of the old "drovers" was the legendary James McKenzie, the sheep stealer after whom the MacKenzie Country is named. There have been numerous accounts of his activities, many of them conflicting, but it seems that he picked up sheep from in behind Fairlie and took them right through to Southland. This would involve crossing the Tekapo, Pukaki and Clutha Rivers – a seemingly impossible job for a man on his own and with only one dog. This dog, an incredibly clever bitch named Friday, was to become no less famous than her master. Whatever the truth of this remarkable droving feat McKenzie was eventually caught heading through the McKenzie country with a mob of sheep stolen from the Levels Station in behind Fairlie. Strangely enough he never had a horse. His swag and gear were loaded on a bullock, which he led, but whether he rode it when crossing the many sizeable rivers does not seem to be known.

One of the great droving achievements of those early days that has been recorded was that of taking sheep from Shag Valley station, outside Palmerston, to Lake Wakatipu when W. G. Rees founded his Queenstown station in 1860. However in the pioneering days when our back country was first being stocked there must have been many notable drives of which nothing is known. Droving in the South Island is now virtually a thing of the past, but it is still commonplace in the north. One of the big North Island drives that is still a regular event is from Gisborne down through Hawke's Bay to Woodville, then through to Ashhurst and up to Hunterville, a drive of some 300 miles.

However even such drives as these are little more than a picnic compared with some of the droving feats of the pioneering days in Australia. I have just finished a book called *Droving Days* by Herbert Barker, a New Zealander who spent many years in the Australian outback; in it he describes various droving feats of the earlier years in that country and it makes fascinating reading.

According to Barker, one of the greatest droving achievements of all time was that of taking cattle right from St George, in south-eastern Queensland, to Glencoe Station 100 miles south of Darwin. This drive, which took a full seven months, was just on 2000 miles. To us in this country that is almost unbelievable. The drover in charge of that drive was Nat Buchanan, who eventually became one of the most famous of the old Australian cattlemen. He was associated with the pioneering of three of the largest cattle stations in Australia – Victoria River Downs, Wave Hill and the Ord. In those days it was impossible to get accurate tallies of the cattle run on such vast holdings but it has been estimated that those three stations, between them, carried in the vicinity of a quarter of a million head.

Needless to add, the horse played a vital part in the running of those giant enterprises. Without them that country could never have been stocked. Herbert Barker, who died here in New Zealand only a year or two ago, went to Australia in the very early years of this century and spent 30 years over there, mainly on station work. He knew many of the old-time drovers personally and in writing of their achievements he gives full credit to the part played by the horse.

Torchy proved to be a good sort of hack; easy to catch, level-headed and quiet, he would go anywhere I asked him. On one occasion at Winterslow I rode him into the shearers' quarters, thinking I was being clever. However my showing off back-fired. Just as he was picking his way gingerly through the doorway one of the musterers sneaked up behind him and jammed a musterer's stick fair up his "arse your mother for sixpence". Torchy gave one bound right into the room. Unable to keep his feet on the bare boards he sprawled to the floor. For what seemed minutes he clawed and scraped and struggled before he finally got to his feet and I was able to pick myself up and lead him quietly outside. I can still hear the laughter and barracking of those ignorant shearers.

Winterslow was then the only station in Canterbury that had no road access to its shearing shed. The whole property, some 30,000 acres, was practically devoid of flat and of necessity the woolshed and the main drafting yards had been built about six miles back from the homestead. Access was then by pack-track only – and of course all the wool had to be brought out by pack-horse. The shed was equipped with a miniature wool press – an intriguing affair – which turned out a bale identical in shape to the normal one, but weighing only 70 to

80 lb. Two of these "pockets", as they were called, just made a load for a pack-horse.

The station ran a five-horse pack-team and with the 6000 sheep clipping about 50,000 lb of wool – in the vicinity of 600 pockets – the packing was a big job. They did two trips a day and with wet weather and the odd hold-up the job took anything up to six weeks. Bill McLaughlin, a Methven man, was the packie when I was there; as well as carting the wool he packed for the musterers and was kept busy for the whole of the season. What he didn't know about packing and pack-horses wasn't worth knowing. The two mainstays of his team – Darby and Joe – were crackers. These two were kept going most of the six months mustering season but the others were required only during the shearing. According to Bill that was long enough too. They were a pretty scratch lot but they did the job. Darby and Joe were as good as any I've seen.

To my knowledge the only other place in this island where wool had to be packed out was at the* Bluff station up the Clarence River in Marlborough. Although the main access into the Bluff was via a 50-mile pack-track from Kekerengu the nearest road was the Waiau-Kaikoura highway at the Clarence Reserve homestead. The track out to the Reserve necessitated fording the Clarence and then climbing up over the Seaward Kaikoura range. It was only about half the distance and it was this route that was used for the bulk of the pack-work. In order to minimise the packing the Bluff sheep were driven out to Kekerengu for the main shearing, but in this class of country there are always several hundred stragglers and these were shorn at the Bluff shed. The straggle wool, together with skins and crutchings and – in the bad old days – countless thousands of rabbit skins, were taken out to the Reserve. At one time the station ran a pack-team of 18 mules and the packing at the Bluff was a major job.

However it was in the North Island that the pack-horse made its greatest impact. In the central districts in particular they did a tremendous job. Murimutu Station – no longer existent – which at one time took in somewhere in the vicinity of 200,000 acres in the Karioi-Waiouru area, once ran a pack-team of 60 mules; at its peak Murimutu shore 40,000 sheep and the whole clip was packed right to Napier, a round trip of 200 miles. All told this would have involved the best part of four months' solid packing.

Huiarau, north of Gisborne, is another of the big North Island stations that ran a big pack-team until comparatively recently. They

* Ben Wethaill worked at this station for three weeks in 1989 (July).

were used mainly for development work – packing fencing material and grass seed – and until well into the 1950s they ran a team of 25 horses. I am told that practically the whole team was saddled up every morning. In 1948, when a fire swept through a big proportion of the property destroying fences and everything in its path, they borrowed horses from neighbouring stations; until the damage was repaired the pack-team totalled 50 horses.

In the North Island in particular the pack-horse played a vital part in the development of our farm lands.

Apart from the packing there was little work for horses at Winterslow. Every inch of the place is "walking" country and when the musterers arrived they simply turned their horses out at the homestead. The cookshop and men's quarters were out at the woolshed and this was the base from which practically all the work was done. All hands, including the shearers, travelled in and out per "Harry Pannell" (walked) and needless to add once out there they stayed put. There was no taking off to the pub every evening.

Except for Darby and Joe the only horse on the place worthy of note was a little grey gelding – said to be half arab – owned by Jack Wolfrey, one of the musterers. Although barely 14 hands it was like a cat on the hill and was as tough as they come. With feet like iron it had never been shod, yet uphill or down, in stoney creek beds and on the hard road, Jack never spared it. It was a great little horse.

When I finished up at Winterslow I headed back home to Ohoka. That ride is perhaps worthy of note. I'd had my saddle sent in to Methven for repairs and as it had not then been returned I had to borrow a station saddle and ride in to pick it up. I had a big day ahead of me so, thinking that I'd get some breakfast at Methven, I left Winterslow at about 5.30 on the Saturday morning. This meant that I was in Methven soon after seven, long before any of the shops were open. As my saddle was ready I decided that I would not wait. With Sheffield my destination that night I planned to travel via the Rakaia Gorge Bridge, through to High Peak Station and Dalethorpe, and out to the West Coast highway at Annat. This would enable me to see some new country, but although it is hard to estimate distances on open country I reckoned that it would be 50 miles. A pretty big day.

High Peak was then managed by Reg White, a well-known Canterbury dog trial man whom I knew well by name. I felt sure it would be no bother getting a feed when I arrived. The thought of it fairly

still virtually just a pack trail, though navigable by vehicles.

made my mouth water. However, my luck was out, for when I finally arrived there just on mid-day the place was deserted. This was a blow as by then the worms were really biting. Even worse was that I'd been counting on getting some directions to help me on the next stretch of my journey.

This was through fenced hill country. Although I knew the general direction, finding gateways could present all sorts of problems. Well that is exactly what did happen; on reaching a fence I'd turn left, looking for a gateway and as sure as hell I should have turned right. If I turned right the gate would be to the left. The boundary fence between High Peak was the worst problem. Sometimes it is possible to tie a fence down sufficiently to allow a horse to scramble over it but on this occasion there wasn't a hope and I had another long ride again before I finally found the gate. The upshot was that it was after ten o'clock at night before I finally arrived at the Sheffield Hotel. How far I rode that day I don't know, but all told I had been 17 hours in the saddle – and without a bite of tucker.

But here again worse was to come. I found the hotel in darkness and it was some time before I was able to rouse anyone out. When the door was finally opened I was confronted by a woman draped in a dressing gown, anything but friendly. I told her my story but got little sympathy – I could have a bed but she flatly refused to get me anything to eat. With no alternative I turned my hack into a paddock out at the back, tied my dogs in an empty shed, and with the old hay-bag showing me where to go climbed wearily up the stairs to my room.

For all the long tiring ride that I'd had the previous day I woke early the next morning – and strange to say I wasn't particularly hungry. The previous evening I'd have eaten a horse. Unable to go to sleep again and still smarting from the landlady's manner, I finally decided that I'd get up and head for Oxford. Sneaking downstairs I let myself out the back door, saddled my horse and stoning my dogs into silence took off down the road. As far as I was concerned that old girl could whistle for the price of my bed.

I was at Oxford about mid-morning and with all shops closed – as well as the pub – I carried straight on to Ohoka, another 20-odd miles. I arrived home about 3 p.m., after just on 48 hours without a bite. It wouldn't be hard to guess how I spent the next half hour.

As for my hack, Torchy, I felt that he had proved his worth on that trip. He'd been turned out when I arrived at Winterslow, and for a horse straight off grass that 80 or 90 miles wasn't a bad effort.

# 5. Coaching and waggoning

MY next job was at Castle Hill on the Arthur's Pass Road, about 18 miles beyond Springfield. It was an easy two-day ride from home and with Torchy still my mount I rode as far as Woodstock the first day. Here, sure enough, I passed my old team still plugging patiently along; little May with her nose a foot in front and that old slug Pryde dragging the chain a foot behind. Apparently she'd quite forgotten the stone under her tail.

The next morning I forded the Waimakariri River straight opposite the homestead. At this point, just where it emerges from the gorge, it is in one stream and appeared to be carrying a fair bit of water. I had my swag lashed to the front of the saddle and it was with some misgivings that I edged Torchy into the stream. This was the first time I'd tried him out in water and a swag tied to a saddle can be dangerous; if it gets caught by the current a horse can be swept from its feet in a flash. Darval Poulton came down to the river to see me over but fortunately it wasn't as deep as I thought and Torchy never faltered.

With the river behind me I headed through Kowhai Bush and out to the main highway at Brooksdale. At the site of the old Porters Pass Hotel – a relic of the old coaching days – a gang of Public Works men, with tip drays, horses and the inevitable wheelbarrow, were working on a new stretch of a roadway overlooking the Kowhai. Before this was put in the river had to be forded three times in about half a mile and in times of flooding could be quite a problem. The new road eliminated two of these crossings. I stopped for a yarn and found that the man in charge was Big Jack Kelly of Springfield. He and his brother Alex were in the contracting game with horses and drays and were well known in those parts. While I was at Castle Hill they took on several contracts, mainly metalling, between Springfield and Arthur's Pass. Jack, a tremendous man standing well over six feet, was a champion on the "banjo" and was also a great man with horses. Alex, Jack and the latter's son, Young Jack, ran three drays, with three horses apiece, and they did a valuable job. Old Jack, who even then was getting on in years, told me that when the Springfield-Arthur's Pass railway line was being constructed he once drove 16 horses carting a bridge cylinder from Kowhai Bush to the Patterson's Creek viaduct. I was later to work under Jack at Brooksdale and he was a wonderful old chap.

There's a story about Jack Kelly that always tickled my fancy. Although a quiet, retiring sort of chap, he once got into an argument with a bush whacker from over Oxford way. The upshot was that this bloke offered Jack out. Seeing that there was nothing else for it Jack agreed, although as they went out he turned to his would-be antagonist, and, remarked, almost apologetically, "I'm afraid you might find me a bit awkward for the first half-hour or so. I haven't had a fight for a good while now and I'm out of practice."

The Oxford man chewed this over and finally decided that discretion might be the better part of valour. I'd say he was dead right too. Jack Kelly was nobody's fool.

At the foot of Porters Pass the road forked, the new road swinging to the right round the sunny side of the spur. The other branch, which even then had deteriorated into little more than a pack-track, was still a good short cut for a hack so I turned Torchy's head that-a-way. This was the original coach road; I am not aware of its grade but it is incredibly steep, and even though all male passengers were made to get out and walk up this stretch it must have been a gruelling haul for those lightweight coach horses. Even more amazing is that this was once the road used by the old waggoners when carting out wool from the stations further up the gorge. That these great cumbersome waggons were ever hauled up that road is almost beyond belief. And bringing them down when loaded with the usual 25 bales of wool must have been a hazardous undertaking. To add to the difficulties, the road is carved round a cold, shady face and during the winter months snow and ice would have been a further hazard.

Beyond the top of the pass the road winds down Starvation Gully, to Lake Lyndon. At the time the old snow poles marking its course were still standing. These poles, which were about 5 ft high, were set right alongside the roadway, every two or three chains. In times of heavy snow, they were all that enabled the coach drivers to follow the track. Starvation Gully – no place was better named – lies into the south-west and with the summit of Porters Pass 3100 ft high it can be a bad place for snow. Castle Hill is some miles beyond the pass and during my first winter there – 1927 – the snow was so bad that we were unable to get out to Springfield for six weeks after one fall. All necessities had to be brought from Cass by pack-horse – a round trip of 36 miles. Although by that time motor traffic was commonplace, many sizeable creeks were still unbridged in the back country and few cars ventured beyond Lake Lyndon. The road from there on was still the three narrow tracks separated by unbroken strips of grass. One track for each wheel and one for the horse. The horses were still indispensable in that country.

My job at Castle Hill had previously been held by a local man, "Snowy" King. He and his brother Ted had been farming at Russells Flat; they were both good horsemen and while in that district did a lot of breaking in. They had, I think, originally come from somewhere over Alford Forest way – The Bush – and the whole family, Walter, Snowy, Ted and Clarrie, were well known in the Canterbury back country. Ted was a bit of a character and there are many amusing stories of his doings. One in particular I cannot resist telling – even at the risk of disapproval. On this occasion he and another local bloke got into an argument outside the latter's house. They eventually came to blows. Hearing the rumpus the other bloke's wife came charging out and joined in the fray. From here on it's Teds story – told in his inimitable drawl, "h"s dropped and all.

"But," said Ted, "She was mostly on 'is side so I hover the fence and took horf, but she was hinto me with boulders and they were bouncing horf me like Hindea rubber balls."

It was at Castle Hill that I had my first experience with a big wool waggon. As with all back country runs it was an essential part of the station's equipment, the Castle Hill waggon still being in regular use. With five lively draughts – two in the shafts and three in the lead – and with the driver perched precariously in his seat high above the right-hand front wheel, they were an impressive sight.

At one time, of course, all back country cartage was done with these waggons and those old waggoners and their horses did a wonderful job. Soon after I went to Castle Hill I got to know the Milliken brothers, who together with their father, John, did practically all the carting for the Waimakariri stations for a quarter of a century or more. John Milliken, who took over Brooksdale Station – which took in country on both sides of Porters Pass sometime about the turn of the century – ran three waggons and became widely known in that part of Canterbury. Two of his sons, Bob and Jim, drove for him for many years, both of them eventually taking on their own properties in that locality. Bob took over the Annadale run at Springfield and Jim went to Flock Hill station which bounds Castle Hill along the Broken River. Jim also held Castle Hill for some years.

Both of them were big, powerful men, ideally suited for that game. Jim, in particular, scaled over 17 stone and was tremendously strong; he used to do his own shoeing and it was said that once he got hold of a horse's leg even the biggest and wildest had no show of getting away. According to one wit the leg would have to come off first. Rugby enthusiasts will remember that one of Jim's sons, Harold, was a 17-stone All Black.

The Millikens' biggest job was carting the Mt White wool. Bob told me that the station was then carrying close on 40,000 sheep. Although the whole clip was scoured on the place, to reduce weight, there were in the vicinity of 700 bales. At one time it was all carted right to Christchurch, 80-odd miles and the round trip took eight days. With the three waggons each taking 25 bales the whole job took well over two months. The old waggon track out from Mt White crosses the Waimak two or three miles above the junction of the Poulter and joins up with the Arthur's Pass highway at Lake Pearson. This cuts about ten miles off the present road, which comes out at Cass, but it involved a big climb from the river, up to the present Midland railway line and over Long Hill to Pearson. Another big climb was up the Craigieburn cutting. The teams comprised the usual five horses, but for any steep pinches this was increased to eight. Three leaders would be unyoked from their own team and hitched to another waggon for that particular climb. The lead horses would then be taken back down and the process repeated.

Horse feed and the men's camping gear were carried on the waggons, and as often as not the waggoners simply unyoked their horses and camped alongside them at night. However all going well this was not necessary, the four stages to Christchurch being Castle Hill, Springfield – where they had a paddock for the horses – West Melton and finally the city itself. The big day was from Mt White to Castle Hill; this is about 26 miles and it was only with a good run that they were able to make it in the day. On this stretch they had to double yoke on four occasions, the worst pinch being the last chain or so up out of the Broken River. The horses' bellies would just about touch the ground getting up there.

By 1907 the railhead had been established at Avoca. As the stations

– Craigieburn, Grasmere and Mt White – then elected to cart their own wool the Millikens took their waggons off the road and concentrated on running their own properties. For over 20 years they had travelled that road, in all weathers, and they and their horses had done a great job.

When I was first at Castle Hill, Jack Cochrane (whose father, Fred, then had the Bealey Hotel) was running an ancient solid-tyred lorry doing cartage jobs from Cass to Springfield. It did most of the station's cartage. I can't remember the make of that truck, but I reckon the horses would have been just about as quick. When going up the Castle Hill Straight, a barely perceptible grade, it got down to about two miles an hour. However it finally ran off the road at Parapet Rock with a load of Flock Hill wool. It finished up in the gully a hundred feet below and that was the finish of that.

The Arthur's Pass road crossed the Waimakariri just above the Bealey Hotel. As there was no bridge then, Fred Cochrane kept a team of horses for the purpose of pulling cars through the river. He had set charges – one pound if you stopped when passing the hotel and asked for a tow; two pounds if you first went and had a look and then came back and got the horses, and three pounds if you had a go, got stuck and then had to get towed out. Fred Cochrane was a renowned fighting man and few argued the point with him. Although there weren't many cars on the road at that time, the horses were a good side-line. There was nothing in the pub then.

Mainly to bolster trade at the hotel, Fred also organised an annual race meeting. This proved to be a popular event. The races – which were open only to men who were working in that district – were held on the Cora Lynn flat a mile or so below the pub, and what with railway workers and station hands there was quite a gathering. Inevitably there was a fair sort of a do at the pub that night and old Fred's strategy paid good dividends.

It was about this time that Harold White, who was then head shepherd at Grasmere, acquired a horse that had met with some success on the racetrack. Naturally he had high hopes of pulling off the big event – the Bealey Cup. That particular year Fred Cochrane had induced a bookmaker to come up for the meeting and knowing that the horse, General Assembly, would be known to this man Harold kept his animal well out of sight until the start of the race. Well, old General duly won that race and the bookie didn't have a very successful day. I might add that Fred Cochrane donated the Bealey Cup himself. That day's racing was the highlight of the year for the people of that comparatively isolated locality.

The Castle Hill waggon was then being used only for odd jobs about the station – carting firewood, carting sheaves when stacking at harvest time, and various odd jobs. The firewood was cut at Enys's Hut, just behind the site of the old hotel and some three miles from the homestead. A contract was let every year for 12 months supply. Although I can't remember how many cords were required, the cartage was quite a job. At 2400 ft the Castle Hill homestead is one of the highest in the country; during the winter months it is the coldest place of my experience and a good supply of firewood is a vital necessity. That particular winter the draught horses were walking about on a foot or more of snow without breaking through the frozen crust. You don't see that very often in this country.

One waggoning job that savoured of earlier days was a trip out over Porters Pass to pick up stud rams. These were brought by motor lorry as far as the first ford of the Kowhai – about two miles from the foot of the pass – and we had to pick them up from there. It was a round trip of some 20 miles. A Kaiapoi man, Alf Clark, was the teamster at the time, and as the shepherd I had to accompany him. It was an interesting trip and even on the "new" road it was quite a haul coming back up over the pass. It's just on two miles from the foot to the top.

Another job which occasionally fell to the Castle Hill team was clearing the road after a fall of snow. The county council left an old horse-drawn grader at the homestead and this was used for that purpose. However after one heavy fall the boss, "Old" Bob Blackley, decided that it would make the job easier if we first of all opened up a track with a "snow-plough". This is simply a V-shaped contraption which, tapering from its apex, opens out to the desired width at the rear end. Normally from 6 to 8 ft in length it opens out to anything from four to 6 ft at its base. With the sides perhaps 30 in high it "furrows" the snow to either side leaving a cleared track the width of the plough. When working in, say, 18 in of snow this will be piled perhaps another foot higher on either side of the track.

With four horses yoked to the plough, and the snow still soft and powdery, we got through to the pass with no trouble. But unfortunately we had a real old man frost that night and the snow, piled high on either side of our track, was frozen like concrete. There wasn't a hope in hell of shifting it with the grader – it would have taken a traction engine to pull it – and all we'd done was mess up the job completely. The catch was that as the snow plough was only 4 ft wide the track was too narrow for any vehicle and with the snow piled high on each side it was weeks before it thawed. Old man Blackley's brain wave back-fired.

Apart from the occasional metalling contract, which was usually let to the Kellys, the maintenance of the road was done by solitary roadmen, each of whom was responsible for a stretch of about 20 miles. They camped in huts, erected alongside the road every ten miles or so and they patrolled their lengths by horse and gig. With their only tools the pick and shovel – plus wheelbarrows which were parked here and there along the road – the maintenance of 20 miles of road by one man seems an impossible job; yet they kept them in remarkable order.

When I first went up that way Bill Bryson, a member of a well-known Springfield family, was doing the Castle Hill length; an ex-musterer, he had the back country in his blood and with only his horse and a dog for company he was as happy as Larry. Bill was followed by Bob Callaghan, a wanderer from who knows where, and I remember him in particular for his wonderful horse Billy. A big brown gelding – trotting-bred I'd say – it was almost human. Old Bob, just by word of mouth, could do anything with him. There were two huts for the roadmen on that length, one at the foot of the pass and one at the site of the old Castle Hill hotel. Many a time when Bob had been celebrating at the Springfield Pub he would climb into the gig, tie the reins to the splashboard and promptly go to sleep. As safe as a church Billy, completely unguided, would take him back to the pass hut, nine miles distant. Needless to add, Bob thought the world of Billy and the horse responded accordingly.

Another, but sadder, memory of Castle Hill is that of one of the station hacks, a lovely little creamy called Nobby. Although I still had Torchy I often rode Nobby, and one day when out on the Torlesse country he came to grief when crossing a little gut. He only fell about six feet, but he landed with a real thump fair on his rump; it appeared that he had damaged his spine, as he had completely lost the use of his hind legs and was just sitting there like a dog. I had no option but to leave him there so I headed for home to report the accident to the boss. Hoping that he may have recovered I went back that night with his cover and a feed of chaff. But there was no improvement and the poor old chap looked a hopeless case to me. The following morning I returned, once again, still hoping for the best, but this time the boss made me take a rifle with the instructions that if there was no improvement I was to put him out of his misery.

As I expected he was still sitting there, powerless to move. So I did the only thing possible. I've had to shoot quite a few horses over the years, but never have I had to steel myself as I did on that occasion. He heard us coming and despite the fact that he must have been in considerable pain he gave a friendly little whinny when I appeared. That didn't help. He was a great little horse, Nobby, and I did it pretty hard pulling the trigger that day.

An incident which demonstrated that the horse was then still more dependable than motor transport occurred one night in the dead of winter. Some motorists heading for Christchurch called in at the Bealey Hotel early in the evening and finding that the publican – a man named Timms – appeared to be seriously ill volunteered to take him to the nearest doctor, at Darfield. However they had the misfortune to get well and truly stuck in Murderers Creek – now known as the Blackball – which is about two miles or so on the Castle Hill side of the Flock Hill homestead. They had no option but to set out and get help. The party consisted of a man and two women, and as neither of the latter relished the idea of being left to look after the sick man it was decided that the two of them, together, would walk to the nearest homestead. However, knowing nothing of the locality, they made the mistake of going straight ahead, instead of going back to Flock Hill – which, of course, they had just passed. This meant that they had a seven-mile walk to Castle Hill, in pitch darkness. When they finally arrived at 11 o'clock they were in a state of exhaustion.

The upshot was that I had to get up, run in the horses and with a couple of draughts head away up the road to pull the car out. By the time we got it out and going again it was well past midnight, and I still had the long traipse back home. How Timms fared I don't know, but if memory serves me right he never returned to the Bealey and it was then – 1927 – that Fred Cochrane took over the hotel.

But it was as the coach route from Springfield to Hokitika that this road was best known. This was before my time in those parts, but an old Springfield identity, Charlie Collyer, used to do odd jobs at Castle Hill, and as he once drove for Cassidy who ran the Cobb and Co.'s coaches for many years he was a mine of information on those almost forgotten days.

Hugh Cassidy, whose huge stables at Springfield were still standing

*Above:* "Access was by pack-track only." Transporting fencing posts by pack-horse to back stations behind Waipiro Bay in the north at the turn of the century.
*Alexander Turnbull Library*

*Below:* At the turn of the century roads were maintained by horse-drawn graders.
*Rangiora Museum*

*Above:* The old cart park in the centre of Rangiora. Note the horses stabled in the buildings to the rear. *Rangiora Museum*

*Below:* Old one-man tip drays used for carting shingle. This photograph dates from pre-World War I days. *Rangiora Museum*

Mt Algidus was still using wool wagons into the early 1950s. These are on the road from Glynn Wye Station to Culverden prior to World War I. *E. M. Topp*

A fine three-horse team. In 1922 Joe Wyllie saw no less than 66 teams operating at once, in the same paddock, at a Rakaia ploughing match. *G. Petrie*

in my time, took over the business in the mid-'70s. Retaining the name of Cobb and Co., he continued until the Otira Tunnel was finally opened in 1923. After the Midland line was put through to Arthur's Pass in 1914, he did only the run over the top to Otira. But at one time his coaches ran right from Christchurch to Hokitika – a distance of just on 160 miles. According to odd reports this was done in two days, with an overnight stop at the Bealey. They left Christchurch at 8 a.m., had lunch at Springfield and arrived at the Bealey somewhere about 7 p.m. Stops for lunch and changes of horses must have taken at least an hour, which meant that the actual travelling time for the 80-odd miles was less than ten hours – an average of about nine miles an hour. With the last half of the journey being through hill country, including the big pull up over Porters Pass, this was great going. There were, of course, several staging depots along the road, where horses were changed, but even so this is quite remarkable.

What is even more amazing is that although Arthur's Pass was not discovered until 1864 – by Arthur Dudley Dobson – a coach with a full complement of passengers actually made the trip right to Hokitika in 1866. One of the passengers, William Wilson of Christchurch, described the journey in a series of articles in the *Press* later that year. The road was then only partially completed and the trip over Arthur's Pass itself must have been a hazardous undertaking. However the construction of that road must have been a priority job as Wilson makes frequent reference to hotels, stores and townships throughout the journey.

Cassidy is reputed to have had over 100 horses when the business was at its height. I am not aware how often they changed horses throughout the trip, but with four and sometimes five in a team a change every 12 miles would necessitate the use of 60 or more horses. With the usual pace a spanking trot this work was particularly hard on horses' legs; they were kept on this job for only a few years. Ample replacements were essential. There was also plenty of other work; feed waggons were constantly on the road picking up chaff provided by various farmers, and this had to be carted to the various staging depots. Big Jim Milliken, who took over Castle Hill when they gave up the waggoning game, provided their needs in that locality. For a number of years Jack Kelly and Jim* Nimmo, a member of another well-known Springfield family, leased land at Riversdale – straight over the Waimak from Cass – for the same purpose. All told the coaching business was a big undertaking.

* Ben Wetherill worked for a Nimmo (Colin) in 1989 on Bluff Station (now called Muzzle Stn.)

Charlie Collyer was driving about the time of the First War and he often mentioned other drivers of that era. Some that I remember are Joe Searle, Sam Eastgate, the Rubb brothers and Denis Clarke. Another well-known Coast driver was Sammy Yde of Turiwhati. When the coaches finally went out of commission Sammy purchased one and kept it as a show piece. I called in there one day and he was polishing it up in readiness to send it over to Christchurch for the centennial celebrations. Sammy Yde drove for Cassidy until the coaches were finally taken off the road.

Another old Springfield identity I often had a yarn with was the blacksmith, George Fraser. George made a speciality of shoeing the coach horses and seemed to know every one of them. Some had to have their shoes specially weighed – to suit their particular gait – and although when I first met him it was many years since he had last shod a coach horse, the walls of his smithy were still lined with shoes of all descriptions which he had kept as patterns. Old George liked to describe in detail the reasons for the weighing and shaping of each particular shoe. Like Charlie Collyer he was wrapped up in those coaching days. Truly that was the day of the horse.

There's a story about old Charlie which is not in any way connected with horses, but it so tickled my fancy that I feel it is worth the telling. At the time Springfield fielded a rugby team – not a bad one either – and one of its staunchest supporters was an old Cornishman. He followed the team everywhere it went and his barrack could be heard all over the grounds. In the course of play there were the inevitable occasions when a player would get bowled over in the midst of a general melee and Jimmy, whose sight was failing, would have to be told what was going on.

"Who darn nah, who darn nah?" he would demand, and if it was one of the Springfield players who was getting the worst of it he would be hot in his defence.

"Let 'em oop, let 'em oop," he'd storm. "Fair play boys, fair play."

But if it happened to be an opposing player who was underneath it was a different matter. Old Jimmy's "Who darn nah" would be followed by an equally emphatic "Kick 'em in the goots boys, kick 'em in the goots!"

Anyway old Jimmy eventually passed on and it so happened that Charlie Collyer was then the local gravedigger. After the burial ceremony was over Charlie went across to fill in the grave and before he started he peered down to bid his old friend farewell. "Who darn nah", he remarked. "Who darn nah."

When I finally moved on from Castle Hill my place was taken by Walter Baynon, who also hailed from the Kaiapoi district. Walter arrived with a horse and gig after having driven all the way from Queenstown where he had last been mustering – a drive of some 300 miles. His aim was to muster – here and there – along the whole length of the island and he eventually achieved his ambition, a year or two later, when he put in a season at Molesworth. He did the whole of his journeying in his gig and although I didn't see much of his horse it was a classy looking bay and must have been a good one.

I might add that just prior to this Walter had taken part in an expedition to the little known Campbell Island. At that time the island was being "farmed" although no-one resided there permanently. The policy was to send over a gang of men, all of whom could both muster and shear, for three or four months every few years. While there they mustered in what sheep they could, doctored and tailed the excess rams, tailed any unmarked ewes, and shore the lot. When their boat returned they loaded the wool and leaving the island to look after itself again returned to the mainland.

Walter told me that when he was there the island was carrying about 2000 sheep, but they were semi-wild and it was impossible to get a clean muster. It had been a great experience, but they had been cursed with miserable weather – persistent drizzle and wind – and they weren't sorry when the job was finished.

I was particularly interested because only the previous year Jock McArthur had tried to interest Darval Poulton and myself into joining him in a scheme to take over the island – lock, stock and barrel. We were both keen on the idea too but for some reason the deal fell through. It's probably just as well, for it's a sure thing that if I'd had anything to do with the proposition it would have been a dead loss. But what an experience it would have been.

# 6. Lake Coleridge days

I PULLED into the Lake Coleridge homestead one morning just as the chaps were running the horses in. It was a sight I've never forgotten. It was in the spring of the year and with the grass almost up to their knees they were just about jumping out of their skins. As they clattered into the stable yard they made a fascinating picture. I was still working at Castle Hill at the time, but the sight of those horses made me decide then and there that my next job would be at Lake Coleridge Station.

As it happened I did go there soon after this, and although that was 50 years ago I remember every one of those horses as though it was only yesterday. The stable, with ten stalls, was split by the harness room, with five horses each side. Each horse, of course, had its own particular stall. At the top end was a bright bay mare, Dolly; a quarter-draught and a lovely level-headed mare. She was the shafter in the pack-team. It is possible to get a spring dray to most of the Coleridge mustering huts and the pack-team consisted of a shafter and two leaders. Dolly as the shafter had an unenviable job, for some of those tracks were pretty rough.

Next to Dolly was a great powerful chestnut known only as The Boss's Hack. This horse, which I understood had once been a police horse, was the 16-stone hunter-type. Although he was a hard mouthed, headstrong brute he had guts to burn and just didn't know the word quit. Jack Murchison, the boss, seldom rode him and he was available for anyone who cared to use him. He was a great horse for a long journey, but was so hard to handle that he got little to do. I borrowed him once but that was enough – he fought the bit all day and my arms had about had it when we got in that night.

Next to him again was a beautiful big iron-grey mare, Bluey. The "Lake" at that time bred their own horses; they had half a dozen or so mares, all fine types, running with a Constad colt, and they left some grand station hacks. Bluey was one of them.

Then came the packies' hack, Ginger, a bright chestnut with a broad white blaze from ears to nose, and four white stockings. There is an old saying about white stockings: "One buy, two try, three doubt, four better without." Ginger belied that completely. As befitting a packies' hack, he was a beautifully mannered, sensible horse. Ideal for that job.

Then in a place of honour next to the harness room was Scotty, one of the "nicest" horses it has been my pleasure to handle. A dapple grey, although then almost white with advancing years, he was the fencer's horse. Old Fitz the fencer, who was many years on that job, did his rounds in a gig and Scotty, as brainy as a man, would follow him along a fence-line all day and never put a foot wrong. To cap it all he was a picture of a horse – a strong cobby type, beautifully put together. Of course old Fitz always had him looking at his best.

Beyond the harness room was the old-timer, Glenthorne. A big brown gelding, well past his best, he had once been the packies' hack. Although still lively enough, he was kept as a spare hack and was seldom used. Then came the other two pack-horses, Roy and Dick. Roy, a bay gelding, was a fine stamp of a horse – a typical pack-horse type – but Dick, a great leggy horse standing about 17 hands, was trotting-bred. Although not as young as he used to be he was a remarkable animal – all fire and guts and go. I used him packing skins when I was shooting deer some years later and although then a big age he was as game and tough as they came. What with green skins and all my gear, I had some pretty hefty loads when shifting camp. Although at times I wondered if he'd ever make it that old bloke never faltered.

As with most older pack-horses you could pile anything on him – with one exception. He objected strongly to carrying my rifle. When heading away for a day's shooting I sometimes took old Dick and tied him somewhere handy to the area I planned to shoot. To save carrying my rifle I'd sling it on the pack saddle. But that old fool of a horse always refused to carry it. On one occasion when he was cutting up rough the rifle came off and landed, point first, in soft ground with the result that I had to go several miles back to camp to clean the dirt out of the barrel. Apart from that Dick was a cracker on that job; I take my hat off to that old bloke.

The last two of the Coleridge string were among the best station hacks I've known. One of them, a sturdy brown cob named Peter, must also have had some trotting blood in him as he was a beautiful ambler. For a "dog-man" this is a wonderful gait; the pace, halfway between a walk and a trot, is ideal for the dogs to follow. A good ambler is a real asset. Peter was a cracker; he just seemed to glide along at about six or seven miles an hour and you just sat there with scarcely a move. At one stage of his life he had gone dirty and no-one would ride him; however one of the shepherds, Joe Casey, who was a master with a horse, had taken him in hand and had made a cracker of him. He eventually became "my" hack and was a beauty.

Lastly was a horse with the commonplace name of Bill. He was owned by the head shepherd, Bob Harrison. A dark brown, a bit over 15 hands, he was bred on the place by the Constad horse, and with the blood of his thoroughbred sire showing in every line he was a classy, attractive horse. Bob Harrison was a hard and fearless rider who baulked at nothing; how that horse kept its feet at times, I wouldn't know. I can still see Bob, full gallop, regardless of the going, with his cry "Ride boldly lads, and never fear the spills." I bought Bill when I left the Lake and he was a lovely horse.

Somehow or other those Lake Coleridge horses have retained a special place in my memories. As a collection of station hacks they'd take a lot of beating. I thought them worthy of a bit of a write-up anyway.

Another Coleridge horse that earned "fame" in that locality was a big, powerful gelding simply known as The Brown Colt. Although a fine looking animal, this horse was an inveterate outlaw, just full of dirt, and as far as I could make out he was kept solely for aspiring riders to try themselves out on. One local man, considered to be pretty hot stuff in the pigskin and as game as they came, bit the dust five times on end one morning before he cried quits. On another occasion a young fellow from Mt Algidus Station rode across just to have a go, but he seemed so young and inexperienced that Bob Harrison would not permit him to mount. As it turned out this chap – his name was Burke if I remember rightly – was a natural-born rider, and he went over to Aussie soon after this and made a real name for himself in rodeo circles. You never know.

Of the local riders the only one to beat The Brown Colt was one of the Coleridge shepherds – the redoubtable Joe Casey. He sometimes hacked him around on his ordinary shepherding work, but the horse was too much trouble and was finally given up as a bad job. He could certainly heave himself around, that horse; today he would have been worth a ton of money to rodeo organisers. Where he came from and how he finished up I don't know – dog tucker would be my guess.

As was usually the case at that time, the musterers arrived under their own steam, either by horse and gig or in the "pig-skin", and their horses were turned out for most of the season. There were four seasonal musterers at Lake Coleridge and their horses also were a somewhat notable collection.

Before I left Castle Hill I had sold Torchy to the teamster, Alf Clarke, who also came from down Kaiapoi way. He had borrowed my saddle and bridle, which I collected later, but by mistake he had taken one of the station bridles instead of mine. Anyway when I finally finished up I simply left mine in its place and it is interesting to note that the "borrowed" bridle is still in my possession – hanging on the garage wall – exactly 50 years later. I might add that it was by no means new when Alf took it and must now be somewhere about 60 years old. Still as serviceable as ever, it is now of no use to me, but – if only for its age – it is one of my prized possessions.

"There's a bridle hanging on the wall."

I'm wondering if I should have it on exhibition as the oldest bridle in the world. At a dollar a time I might make good money.

Before I set out for Lake Coleridge I had a trip to Tattersalls and bought myself a new hack – the first of several I was to buy at that famous old horse sale. A chestnut mare, said to be by Peter Rita, she was a straight out trotter – one of those slow ones that seem to abound. However as it happened she did once win a race for me. And, I might add, she was up against some of the best-bred horses in the country. The event was over half a mile, the track was the hard road in front of the Coleridge cookhouse, and the spectators were the shearing gang and a few permanent station hands. The stake was £4 – £1 from each of the four starters. There was no tote, but old Charlie Smith, the butcher and brander, was running a book.

The starters, all highly-bred were: J. McIver's Denver, a chestnut gelding by Denver Huon; F. Bushell's Lucy, a brown mare by Man O' War; P. Newton's Rita, a chestnut mare by Peter Rita; and J. Harmer's Homeward Bound, a brown gelding by What's His Name. Blood to burn and all rearing to go.

Well, P. Newton's Rita won that race hands down. Denver, a pacing horse, proved to be all action and no progress, Homeward broke badly at the start and Lucy was certainly no credit to her famous sire. After passing the winning post about two chains ahead of the next horse I had thoughts of taking the racing game on seriously, but when I heard what her time had been I dropped that idea smartly. However had Homeward settled down and got going he'd have left my mare standing. He was a slashing gig horse with tons of guts and he could really move. Once, in response to an urgent call, he did the 30 miles from the Lake homestead to Glentunnel, in an ordinary gig, in two hours and ten minutes. As a final effort he did the last mile in three minutes. Given his head he'd go until he dropped. Old Homeward Bound was a real horse.

For all his unimpressive display in the "Lake Coleridge Stakes" Denver was also a great harness horse and a lovely hack. Although trotting-bred horses are not generally the best on their feet in rough going he completely gave this the lie. When cantering over uneven tussock he was like a feather – safe as a church and beautiful to sit on.

Actually Rita – that's too flash a name for a musterer's hack anyway – turned out to be a bit of a plaster. One morning when she'd had a week or so's spell I went out to get her, and having taken a bridle I hopped on her back intending to ride her back to the stable. But Rita had other ideas. For no apparent reason she suddenly put her head down and really went to town; she headed for some willows, with branches almost touching the ground, and I remember wondering how I'd fare when we reached them. But I needn't have worried – I bit the dust well before that. The really annoying part was that when I climbed on her there wasn't a person in sight, yet when I picked myself up just about everyone on the place was there to cheer me.

The strange part was that prior to this she had never shown the least sign of dirt, but from then on every time she had a few day's spell she'd have a go. Her victory that day must have gone to her head. Fortunately I was able to sell her to a bloke who just loved a rough horse. It was then that I got Peter to ride so all ended well.

Although it was in no way due to my big win with Rita, it was while at Lake Coleridge that I first became involved in the racing game. To tell the truth I had little option. I found that among back countrymen, especially up the Rakaia, it was traditional that one should attend both the New Zealand Trotting Cup in November and the Grand National Steeplechase in August. One just had to go. Indeed I recall one old bloke at Lake Coleridge – who obviously didn't know one end of a horse from the other – being asked if he was going down to the Oaks.

"Oh my goodness," he replied. "I certainly can't miss the Oaks."

Like me he'd never even heard of the Oaks before that, but he was working up the Rakaia and heaven forbid that such ignorance should ever be known. Anyway down to the Oaks he went, and by all accounts he came back a sadder and wiser man.

As for myself I also fell in line, although I can't say that doing my money cold in slow racehorses was my idea of a hobby. However I was soon discussing the merits of such well-known horses as Ahuriri, Peter Bingen, Harold Logan, Oakleigh and Willshire with such authority that you'd have sworn I was born in the same stable. However I regret to say that with one exception I invariably backed the slow ones. That one exception was the mighty Indianapolis, who won the New Zealand Trotting Cup on three occasions. I backed him each time and was enthusiastic enough to put my money on him for a fourth win. He made a great attempt, but the gallant old fellow couldn't make it.

It was at Lake Coleridge that I first caught up with the riding swagman, Spud Murphy. A big bearded man, hailing from Southland I believe, he did a round of the South Island stations every year and called in at the Lake as regular as clockwork. When I knew him he was riding a pretty classy chestnut mare and he invariably had a foal which followed along as quiet as a pet lamb. Without as much as a beg your pardon he'd pull into the stable yard, give his horses a good feed and then head up to the cookhouse and practically demand a meal for himself.

He struck a pretty crusty old cook on one occasion and got given his pedigree in no uncertain terms. Old Spud listened, almost in boredom, until the "babbler" ran out of breath; he then turned and calmly looked him up and down for a full minute.

"I presume you're the cook," he finally remarked. "Well let me tell you, little fellow, I've had a few feeds on this place before you came and I'll have a few more after you've gone."

One year he even complained to the boss about the poor quality of the chaff – and got away with it too.

I never saw Spud Murphy after I left the Lake and often wondered how the old chap finished up. He was an independent old so-and-so, and his horses were just about everything to him.

I have many amusing memories of those Coleridge horses. Yoking the pack team into the dray at the start of the season always provided a bit of fun and excitement. Although normally no trouble, those horses were as mad as maggots after the winter spell and they always put on a performance that could have put some of those rodeo horses to shame. First of all Dolly, the shafter, would be yoked up. While one man held her head the two leaders, a man to each, would be backed into position. The driver would then be handed the reins and with one man to each set, the leading chains would be ever so carefully hooked to the Ds on the shafts.

The click of the chains was enough. As the five assistants leaped to safety those mad horses took off flat to the boards. With Bob Harrison the driver, as of right, they would disappear down the road in a cloud of dust, the empty dray bouncing like a yo-yo and Bob in his glory. However that one mad scamper was enough; from then on they behaved like ordinary horses, although it was always reckoned that no-one could drive that team unless he had a ticket.

There was another occasion when we wouldn't have minded a bit of action. With some little job to do we had Dolly in the dray on her own when for some unknown reason she refused duty. The dray, which was empty at the time, was no weight but she just dug her toes in and refused to move. This was about mid-morning, so after we'd tried all we knew to get her going – I even suggested the old hot-stone-under-the-tail trick – we decided to hook the reins up, tie the wheel and leave her until after dinner. When we came back it was the same – never a budge. So once again we tried the same treatment, this time leaving her until knock-off time at five. But that bloody stupid mare still wouldn't move an inch. So, not to be beaten, we left her tied like that all night – that was just on 24 hours all told.

That fixed her; she made no bones about moving off next morning and as far as I know she never jibbed again.

On another occasion we were riding back to the homestead one stinking hot afternoon and on reaching little Lake Georgina one of the chaps suggested that we take the horses in for a swim. I happened to be riding old Scotty, and I stripped off and followed them in. I'd often heard it said that horses are natural swimmers, but as far as Scotty was concerned that definitely wasn't the case. The other two blokes who went in ahead of me had their horses swimming round like ducks, but all old Scotty would do was make great plunges. Each plunge took him out another yard or so into still deeper water until finally we were going almost out of sight. But he still made no attempt to swim; down he'd go, hit bottom and with a tremendous plunge shoot almost clean out of the water. Then down to the bottom again and with the depth always increasing. For a start I thought this great fun but as we got further out the old porpoise trick lost its appeal. I baled out and left old Scotty to it. I'm no swimmer but I managed to thrash my way to the shore.

I was too intent on my own struggle for survival to worry about Scotty, but he eventually found his way out none the worse for wear. But that episode cured me of swimming on horses. A wee bit of water is alright now and again, for washing in or putting in whisky, but I don't like it in big quantities.

However there was another occasion when I saw a real exhibition of swimming. This particular morning we were going out to muster the Peak Hill paddocks. To get to this block we had to follow round a narrow bridle track overlooking Lake Coleridge itself, about eight feet above the water's edge. One of the chaps was leading his horse – a beautiful big bay mare he thought the world of – when she slipped and, unable to regain her feet, landed down on a little beach bordering the foot of the hill. There was no hope of getting her back up just there, but with the beach in only a foot or so of water it would have been possible to ride her round, hard against the bank until it became less steep.

But Len didn't fancy this and as I'd been having a hell of a lot to say he challenged me to have a go myself. That didn't appeal to me either but after all my talk I could hardly refuse. Anyway I scrambled down, caught the mare and climbed aboard. But when I dug her in the ribs she immediately backed straight into the lake. With the beach dropping steeply she was out of her depth in no time. Regardless of the fact that I was in full marching order I left her to it and floundered ashore – wet to the hide. Needless to add this brought a great cheer from all hands, but I thanked my lucky stars I had for that mare turned and struck out straight across the lake for home. It must have been about 300 yards to the far beach, but she did it like a champion. Whether she'd have made it with me on her back I don't know. That could easily have been Newton's last ride.

While at Lake Coleridge I used to have a break down country during the winter – digging spuds or something equally idiotic – and having sold Rita I decided to get myself another hack. So down to Tatts I went – and once again I bought a plaster. With old Jock McGregor's whip snapping round its hocks it pranced around the ring like a two-year-old and looked good. But, as I was once told by a bloke who had rushed into marriage, you've got to have more than looks.

This horse, Norman, proved to be just about on a par with that old Woodstock slug Pryde. He'd have been just the job for those Haywards drivers – trotting slower than a walk would have been right up his alley. Actually I only once knew him to show a turn of speed and on that occasion he was like greased lightning. At the time I was filling in the winter kea shooting for a crust out on the Glenthorne country. I had a "call" bird (a captured kea) which I used to take out with me and tether in some likely place in the hope that it would call other birds to me. Anyway I was shifting camp one day and had this bird, his cage inside a sack, hanging from the side of the saddle. Unbeknown to me it had torn a hole in the sack, for suddenly with a piercing cry it sank its powerful beak into Norman's ribs.

From a standing start that horse hit top speed within five yards, and how I stopped with him I don't know. We did a mile and a half before I even got him eased up and I reckoned we did it in two and a half minutes dead – two and a quarter seconds better than Rapier had recorded in winning the Wellington Cup (over the same distance) the previous year. No-one could have called Norman a slug that day. But that was the one and only time he ever showed the least sign of life. It was also the only time that that bloody "call" bird ever opened its mouth. It just used to laze in the sun and watch me while I did the calling. Both the bird and Norman were a dead loss.

In those days it was almost an unwritten law – at Lake Coleridge anyway – that on arriving at the station for the start of the season each of the musterers should produce a bottle of whisky. It also seemed the custom to drink about half a bottle before leaving the last pub and there were some fairly hectic drives as those wild men headed up country. On one occasion two men travelling per horse and gig were seen heading up the gorge full gallop. The driver, as full as a bull, was sitting on the front board with a foot on each shaft and belting hell out of his horse with the reins. His mate blissfully asleep was curled up in the "basket" slung above the axle.

Another hard case, travelling per pig-skin, couldn't resist having a swig out of his bottle as he went. Careful not to spill any he stuck his finger down the neck of the bottle between drinks, but somewhere along the line he must have fallen off his horse and the bottle was smashed to smithereens. Undaunted, he managed to climb aboard again and finally finished his journey. All he had as evidence that he had intended to shout for his mates was the neck of the bottle still firmly attached to his finger.

Mt Algidus – made famous by Mona Anderson's writing – adjoins the Lake Coleridge country along the Wilberforce. Of all the South Island stations it would, I think, be the last to have used the old wool waggons. They were in use carting the clip until about 1952. By that time the road had been extended round the foot of Mt Oakden to its present terminus straight across the river from the Algidus homestead. The station then purchased a big four-wheel drive army truck, which replaced the horses and waggons. Prior to that they had an eight-mile drag out over the river to the famous old Iron Store on the flat below the Peak Hill homestead. Algidus was then shearing about 14,000 sheep and with the clip totalling 350 bales or more the cartage of the wool was a big job.

The Wilberforce is one of the most treacherous rivers in the island, and over the years it has claimed the lives of a few horses – and men. It was while I was at Lake Coleridge that two trampers were drowned in that normally harmless looking river. Fifteen mounted men combed its streams for several days but one of the bodies was never found.

Indeed we were lucky that there wasn't another casualty. One of the searchers, mounted on The Boss's Hack, floundered into a bad patch of quicksand and it was only that horse's tremendous strength and fighting spirit that enabled them to struggle back onto dry ground. There was no criticism of The Boss's Hack that day. He was a mighty horse.

At that time Mr Algidus must also have been the only place on which a horse-powered chaffcutter was still operating. This "machine" consisting of four poles extending horizontally for six feet or so from a central hub (like the spokes of a rimless wheel) which, by a system of cogs, rotated a shaft that was connected to the cutter. The horses, each of which was yoked to one of the poles and coupled to the one ahead of it, simply walked round and round. This supplied the power. This machine was actually in use at Algidus until about 1930.

I've many good memories of those days up the Rakaia – known simply as "The Gorge" – and not least of them were these experiences with the horses. But with the desire to see new country I decided to once again move on. I had managed to find a buyer for Norman – believe it or not – and in a weak moment Bob Harrison had agreed to sell me his hack Bill. At last I had a real horse, and it was not without a feeling of pride that I mounted him one sunny morning and waved "The Lake" goodbye.

# 7. Billy the first

THE next season, with my compass set further afield, I headed for Mesopotamia station up the Rangitata Gorge. My new hack, Bill, was a beauty – a lovely horse to handle, a great walker and best of all a beautiful horse to sit on. I was looking forward to the long ride. I planned to travel via my old home, Woodstock, cross the Waimak and go through to Lake Coleridge via Porters Pass and Lake Lyndon. From there I'd cross over the Rakaia, follow up that river as far as Lake Stream and then cut through to Hakatere and Mt Potts, which is straight across the Rangitata from Meso. This would enable me to see some new country, and it was also the shortest route.

But it was not to be. The first day I made Woodstock alright – a ride of 30-odd miles – but there had been heavy rain in the back country and the river was pretty high. Too high for me. As I still intended to go through to Lake Coleridge it meant that I had to make a detour back down to the gorge bridge and up to Springfield, an extra 20 miles. From Springfield, of course, I just carried on over Porters Pass, a ride all told of over 40 miles. However when I arrived at the "Lake" I was told that the Rakaia was a banker, so once again I had to change my plans. This left me no alternative but to head down to the Rakaia Gorge bridge and head through to Mt Somers.

Phil Green, whom I'd mustered with at Winterslow, was also booked to go to Meso. As his home was at Alford Forest I decided that I'd join up with him. From the Lake to the "Bush" is just over 30 miles so that filled in another day.

Phil was travelling by gig, so with Bill hooked on behind we set off for Mt Somers the following morning. The usual route from Mt Somers to Meso is up the Ashburton Gorge, past Hakatere station and through to Mt Potts. But this meant fording the Rangitata. Phil, who knows these parts well, ruled that out, deciding instead to stick to the main road, cross the Rangitata at the Arundel bridge and park up at Mt Peel that night. This meant another 30 miles, but when we saw the river we were damned pleased we had gone that way. It was as thick as pea soup and carrying a lot of water. The next day we had only to follow the road right to Meso. This was also about 30 miles, making the total distance I'd travelled 160 miles or more.

As there is a bit of hill work for the first few miles past the Mt Peel homestead I decided I'd ride for the rest of the journey, and Phil set off on his own. I never caught him up all day. His horse, an unusual sort of dun colour, was a good 'un and as any experienced horseman will know a horse in a gig will always out-pace one that is ridden. I thought Bill would have footed it with him, easily, but Phil left me far behind.

At that time the Aclands, who have had Mt Peel ever since it was first taken up, employed a manager, David Pringle. As he was riding up the road himself that morning I joined him for the first few miles. He had a lovely hack, a big grey gelding which was reckoned to be the best walking horse in that locality. But although I say it myself, Bill had his measure. There was a real gorge wind blowing down the river that morning and one sudden gust sent my hat flying. By the time I retrieved it Pringle on his big grey was several chains ahead, but Bill, at a straight walk, overtook him in no time. Davey Pringle was all eyes and ears after that. But he had a beautiful horse all the same.

Mt Peel was one of the few places that used mules in their pack-team. At one time they kept a jack donkey and bred their own. The station originally took in both sides of the range right to Bullers Saddle on Ben McLeod, and with a big proportion of their country on the far side of the range packing was a major job. On the Rangitata side the country rises as steep as a hen's face and it's a long, hard pull up over the top for a pack-horse. The mules were tough and hardy in the extreme and the Mt Peel people swore by them, for all their vagaries.

However there was one Mt Peel man who didn't quite hold that view. One day when this man, Bill Carney, was doing his rounds he spotted that the door of one of the musterers' huts had been left open. He went across to investigate. He found that one of the mules had found its way inside and was contentedly chewing away at one of the straw mattresses. Thinking to give it a lesson Bill picked up a length of firewood and sneaking quietly in past the mule he gave it a good crack across the rump. In its startled jump the mule slipped and knocked the door shut – and then the fun started. One thing a mule can do is kick, and according to Bill this one was in world class. There was a row of bunks two high round the walls of the hut and that mule had Bill skipping round those top bunks in a way that would have put Olga Korbutt to shame.

By the time the mule had exhausted itself – hours later according to Bill – the bunks were smashed to matchwood and it was a very civil man who cautiously sneaked round the ruin of bunks and opened the door. I worked with Bill Carney about 20 years after that episode and he still hated mules.

My ride up to Mesopotamia that day took me past the Ben McLeod homestead, and it was here that I first met Herbert Barker – whom I mentioned in chapter four. During his time in Australia Barker spent some years doing contract cartage work with both camels and donkeys. He wrote a second book, *Camels in the Outback,* in which he describes his experiences while on that work. One of the illustrations is of 30 donkeys hauling a load of timber, estimated to be over ten tons. Another interesting picture shows a bull camel packing four full-sized bales of wool – a load of well over half a ton. As a comparison a pack-load for a horse seldom exceeds 200 lb.

Herbert Barker never lost his fondness for donkeys, and after coming back to New Zealand he always ran a few at Ben McLeod. They were great little animals and provided a lot of fun. Various musterers tried their hand at riding them, but those donks had too many tricks. They didn't actually buck but simply stuck their heads down between their legs and with a series of shuffling, humping movements gradually worked their riders forward over their withers. When the time was right an extra little jolt had their riders standing on their heads. The Mesopotamia musterers used to help out at Ben McLeod and one of them, Jimmie Binnie, had a go one day. Jim scaled about 14 stone – about the same weight as the donkey – but the story was the same. All laughter, Jim was on his hands and knees, about to get up, when the donkey wheeled and gave him both barrels fair in the behind. All hands had to climb out over the top the next morning to muster the back of the range, and Jim could hardly walk.

Mesopotamia is, of course, "walking" country. Consequently horses do not play a big part in the station's activities. However at that time they had a collection of beautiful "ponies". As I remember D. G. Wright, then of Surrey Hills, down Mayfield way (he had a financial interest in Meso), bred Welsh cobs; several mares, by a Welsh cob sire, had been set up at the station. There were four of them; a roan, a grey and two bays. Although only about 14 hands, they were sturdy, beautifully coupled horses, ideal for hill work. Jimmy Binnie, one of

*Above:* A well-known overnight resting place on the coach route from Springfield to Hokitika. Coaches outside the old Bealey Hotel. *Canterbury Museum*

*Below:* "Lord of all he surveyed." "Shiny Bob" Allen's hansom cab became almost an institution in Christchurch. This photograph was taken in 1935. *Canterbury Museum*

*Left:* A photograph over a century old. Eight wool waggons coming out over the Pig-Root, Central Otago. Taken in the 1860s. *Hocken Library*

*Below:* There are no fewer than 112 working horses lined up here at the beginning of the day on the old Springfield Estate at Methven. *Canterbury Museum*

the musterers who was a pretty good doer – to put it politely – always rode the roaney and it carried him up hill and down without turning a hair. Saddle and all, that little mare would be carrying all of 15 stone and she did it with ease.

During the season Phil Green decided he'd break the grey to harness, but although she was a lovely level-headed mare to hack she really went to town in the shafts; Phil had all the fun and excitement he wanted. However she finally settled down and made a good gig horse. At that time the mail and odd groceries had to be picked up at Ben McLeod, six miles distant, and this was usually a job for the gig.

Another horse on the place that really earned a reputation for itself was a brown mare belonging to Bill Hardie, the head shepherd. She had been acquired from a travelling buck-jumping show when Bill was at Mt Algidus, and when it came to swopping ends I've seen few who were her equal. One nippy sort of a morning we were heading out to muster the Downs and Bill decided he'd take this mare. He was wearing an oilskin – not the best of things with a touchy horse – and sure enough that mare really got stuck into it. Bill stayed with her, but when it was all over he felt in his coat pocket for his tobacco packet and what should he fish out but a handful of horse dung. I knew she went high that mare, but hell!

It might take some believing, that story. "But it's true, and if you don't believe it, well, you know what you can do!"

When being shod that mare also put up a great fight. She had to be thrown and roped, and even when trussed up like a fowl she fought and struggled until the last nail was driven. She was just a born fighter, but what a horse she could have been.

There was one performance however that didn't finish so well. That mare was too many for Bill this time, with the result that he had to take to his cot for a few days. After that fall he decided that she was a dog tucker job, but before we skittled her we took her down to the stockyard, strapped on an empty pack saddle and let her go. And she *did* go. Without the weight of a rider to hold her down she went so high that two or three times she somersaulted clean head over heels to land smack on the broad of her back. I've never seen a horse get its hind end so high. Its no wonder Bill fished dung out of his pocket that day.

I might add that we concluded that the dung must have been flung up by the mare's heels during the performance. But how it landed in his coat pocket I wouldn't know.

With a good scope of river flats, the station ran a six-horse team and as well as growing a substantial area of turnips there was always a paddock of oats for horse feed. The oats, of course, were cut for chaff and although I was there in the 1930-31 season, when portable chaff cutters similar to those in use today were commonplace, the Mesopotamia contraption was little better than the Mt Algidus monstrosity. Certainly it was driven by an engine, but there are engines and engines and as far as that one was concerned I'd have backed the horses every time.

Actually the chaff cutting, which seemed to be almost a fortnightly job, was quite an experience. No-one looked forward to it, but it certainly was a change and was always good for a few laughs. The engine – a single-banger Hornby (if that means anything) – was a fixture and as such was exposed to all weathers, winter and summer. Consequently it was practically impossible to start the bloody thing. Fortunately there was a man on the permanent staff who had a cousin who married a mechanic, and naturally there was nothing about engines that he didn't know – or so I gathered. All the same it appeared to me that he thought that you could work engines the same as you could a dog. He used the same sort of language anyway. It always took about an hour to get it started and you could bet it would fizzle out about every 20 minutes. Not that that worried us.

The chaff cutter itself, which was also a fixture, was only about 3 ft high, and from the cutting blades the chaff simply dropped onto a boarded shute at ground level. The bag-hole actually was a hole in the ground. About 3 ft deep and 2 ft in diameter – just big enough to accommodate a chaff bag – it had an iron framework round the top from which a sack was attached so that it hung in the hole with the bottom just touching the ground. The catch was that one man had to stand in the bag and with a wooden scraper rake the chaff into it. As he did this he had to tramp the stuff as tight as possible. This was done almost entirely by feel, as after about ten seconds the man was practically blinded by dust. Everyone bar the teamster took his turn in the bag hole, so at least we had the opportunity to laugh at each other.

It was a job that was not keenly sought after; indeed, I've dug spuds in heavy clay country, shod outlaw horses, dagged fly-blown ewes in a wet spring, milked cows with sore teats and cut manuka scrub after a fire has been through it (you try that) and any of those jobs were preferable to working in the bag-hole on that Mesopotamia chaff-cutter.

However there was one saving feature. The engine kept conking out and there were plenty of spells. The sheaves had to be carted by dray from the nearest stack and it was seldom that more than one drayload – 30-odd bags – was cut in a day. The horses had the best job that day.

And there was another job that savoured of the old pioneering days. At that time there was a sizeable island out in the Rangitata River bed and as it was lovely sweet country and reasonably handy to the homested, the rams were wintered there. As the island was cut off by a stream of some size the rams had to be carted across in the old farm dray; there were a hundred or so rams and as the dray would take only a dozen at a time it was fairly slow progress. This was in 1931, yet here we were doing the job in exactly the same way it would have been done in Samuel Butler's time – nearly three-quarters of a century ago. The only difference was that Butler would have used bullocks instead of horses. Samuel Butler, of course, founded Mesopotamia station in 1860.

When the mustering season finished at the end of May I headed down country to spend a winter on the "mud flats". I was showing some signs of being a useful rugby player – just useful enough to fill a gap if no one else was available – so I joined up with the Ohoka Club. They were always scratching to field a team so I was always sure of a game. By this time the country was in the grip of the great depression. I couldn't even get a job digging spuds. However I enjoyed the football – and eating the old man's tucker – and I was able to spend the whole week working out blinding moves for the following Saturday's game. One of those moves nearly came off once; if it had I reckon I'd have been hailed as another Bert Cook.

With sheep being sold at almost give-away prices, I got the idea of trying to get the grazing rights of the northern side of the Waimakariri River bed – right from East Eyreton to the gorge bridge. I discussed it with my old boss R. O. Dixon and although there would be a certain flood risk he was all for it. The idea was that I would buy a line of ewe hoggets – with R.O. finding the money, hopefully – and turn them over the following year as two-tooth ewes. However it would first of all be necessary to make a comprehensive tour of the area to assess the possibilities.

This would have to be done on horseback, but with Bill eating his head off this would be a good opportunity to make him earn his tucker. There was only one problem – the distance. I could only guess at this but it would obviously be a tremendous ride to do in one day. The stretch of river bed alone would be the best part of 25 miles, and as the Waimak itself was about ten miles from our place the ride to the gorge bridge in a direct line would be over 30. The extra riding involved in zig-zagging about the river bed, inspecting the various islands, was anyone's guess. On top of that there was a 30-mile ride, by road, back to Ohoka. At least 70 miles, all told.

An alternative was to take my sleeping bag, a rope to tether Bill, and a bit of tucker, and spend the night parked under a gorse bush somewhere along the way. This didn't appeal to me and I decided that I'd do it in the one day, regardless.

Well, my estimate of the distance can't have been far out as that ride took just on 18 hours. I left home at four in the morning and it was just on 10 o'clock at night when I got back. I had to travel at a walk most of the way, but Bill was a great walker and we must surely have averaged four miles an hour. Yet Bill finished that day in great fettle. On the way back we simply stuck to the road and for the last few miles, when the country became familiar to him, he kept swinging into a canter on his own accord. That horse surely proved himself that day. Seventy miles is a big day for a mounted horse, and on top of that a third of the going was over rough, stoney river bed.

I might add that that scheme of mine came to nothing. Too many of the landowners adjoining the river bed were opposed to the idea and it wouldn't have worked out. So I stuck to the mustering.

I headed north to St James Station the next season. With memories of the Rangiora horse sales and my little favourite, May, of the Woodstock team, it was their horses that was one of the main attractions. St James, some 200,000 acres, extends from the Captain Range, which overlooks Hanmer Springs, right to the main divide – it takes in some wild, broken country. Although at that time it did have road access from Hanmer to the homestead – a distance of 20 miles – it involved a big pull up over Jollies Pass and except for the carting of wool at shearing time it was seldom used. A pack-track had been opened up over Jacks Pass and as this cut the distance to only eight miles practically all the "cartage" was done by pack-horse. The St James pack-team was a common sight in the Hanmer township.

Owing to the difficulty of road access the St James homestead was built alongside the Clarence River, right in the south-east corner of the run. This is a cold, bleak spot and as the best of the country is over in the upper Waiau, beyond another high range, this is where most of the station work is done. An out-station, known as the Ada, was established in the forks of the Ada and Waiau Rivers, 24 miles back from the homestead. As this was only accessible by pack-track, packing has always been a major activity in the running of the station. A feature of the Ada is the great stretch of river flats. These are mainly high quality river silt, feet deep in places, and at one time extensive areas had been cultivated. This meant that all the agricultural machinery had to be taken out on pack-horses. To do this everything had to be dismantled – and re-assembled when they got it out there – and as there was a fairly extensive range of plant this was quite an undertaking. Those old pack-men must have had a few clues. I thought that packing a live boar wasn't a bad effort, but this would have been child's play to those old blokes. If they wanted something they got it there – somehow.

The most unusual packing job I know of was that of taking a piano into Ngatapa Station up the Mohaka River in northern Hawke's Bay. It was slung on long poles between two horses, and they tell it didn't even need tuning when it got there.

I first knew the St James horses in the early '20s when they were a feature of the Rangiora horse sales, and they were then a distinct type. Predominantly a dark brown they were low set, nuggety horses, ideal for pack-work. Although a bit on the small side they were in great demand as team horses. The station was then held by C. H. Ensor, and it seems that he was then using lighter sires. Ensor sold to James Stevenson in 1927. Reg Barnes, who went there as a stockman about that time, tells me that their main sire was a Masterpiece horse.

However Stevenson, who lived at his Flaxton property – just outside Rangiora – bred Clydesdales. Deciding that the St James horses were too light, he started using Clydesdale stallions. These, of course, left a bigger type of horse and it was probably from their progeny that the St James horses developed the jumping ability for which they have since became renowned.

I mustered on St James in the early '30s – in the Flying gang – but as we did only the hill work I didn't have much to do with the run horses. The brood mares, somewhere about 70, were run at Stanley Vale, a beautiful valley on the homestead side of Lake Guyon. Stanley Vale was once a separate holding, but was amalgamated with St James well before the turn of the century. The balance of the horses were run out at the Ada and had the run of practically the whole of the upper

Waiau Valley. They did particularly well on those river flats and it was a great sight to see a mob of them in full flight when being mustered. The station was then running somewhere about 200 horses, all told.

Although we musterers didn't normally take part in the handling of the run horses there was usually quite a collection in at the homestead, and any aspiring young rider had plenty of opportunity to try himself out. Some of those horses were pretty obliging.

I had a definite leaning for the quieter ones myself, and two that I remember in particular were a little bay named Teddy and a black which went by the venerable title of Daddy. With names like that they couldn't be anything but quiet. Teddy was the cook's horse, and judging by the heaps of horse dung in the vicinity he seemed to spend most of his life just grazing in the kitchen window. The cook at that time was a bloke named Hank, and as was the practice on most stations in those days his duties included the making of bread. To say the least his prowess as a baker left much to be desired and Teddy's main job seemed to be eating the dud batches. Fortunately he wasn't as fussy about his tucker as the station hands.

Hank was an enthusiastic – if not very competent – photographer. At any hour of the day, even if halfway through making a batch of bread, it would suddenly occur to him that the light conditions were dead right for some picture he had in mind. He would down tools, clap a saddle on Teddy, grab his camera and be away at a gallop. Teddy didn't mind as he was pretty sure of a good feed of bread when that particular batch came out of the oven. Unfortunately Hank was no better at using a camera than he was at baking bread. I never saw anyone else ride Teddy.

Although getting on in years, Daddy was a lovely little hack. He was beautifully quiet and level-headed, and the very greenest and most inexperienced of riders could take him. He'd never put a foot wrong; he was a real asset to the station. Although the road over Jollies Pass was kept in reasonable order no motor vehicles were kept in at the homestead. Any new arrivals such as cooks, shed hands and the odd rouse-about were picked up at Hanmer and taken in via Jacks Pass. Hacks would be taken out for them and old Daddy was kept especially for this purpose. Some of those blokes had never before been on a horse's back.

With Hanmer Springs only eight miles distant the station hands were regularly riding out to various "dos", and while the trip out presented no problems this was not always the case when coming home. On one occasion we were returning after a particularly heavy all-night session. As we were making our way up the pass a ghostly apparition suddenly appeared a chain or so below the track. There was a threshing of bushes accompanied, as we got closer, by a string of muffled curses. This, of course, ruled out ghosts (it's well known that ghosts never swear), so we went down to investigate.

We found that our ghost was nothing but a white horse. Its rider, who was heading back home as full as a bull, had dropped off to sleep and fallen off. Too drunk to climb on board again on the narrow pack-track, he'd led his horse down to a little flat. There was a sizeable manuka bush there which he reckoned he could use as a ladder to get back on his horse. However the main "trunk" of the bush, which was only about as thick as a shovel handle, wasn't strong enough to hold him and his scheme just wasn't working out. How long he'd been there we didn't know – he didn't either, but he reckoned it must have been all of two hours. Anyway when we arrived that manuka bush was trampled almost into dust; all that remained was a solitary spike about a foot high and that bloke, a trier if ever there was one, was still endeavouring to climb it. That white horse must have been almost as patient as old Daddy.

Well, we got him mounted and hit up over the pass, but when we dropped down into the Clarence flats he challenged all hands to a race home. With a wild yell he sunk his spurs into his horse's flanks. She gave one almighty bound – leaving her rider sprawling on the ground – and disappeared in the darkness flat to the boards.

At the end of that season I decided to sell my hack, Bill, although I did it hard parting with him. He was a lovely horse, but we were then in the dead of the slump and I was just too lousy to buy feed for him through the winter. A season's mustering was only worth about £60 then and a chap had nothing to spare. I sold him to Hanmer Atkinson, the owner of Woodbank station – one of the places that the flying gang mustered – and I did so with every confidence. However I heard later that Bill fell with him about the first time he rode him and that old Hanmer condemned him out of hand. This was beyond me. I'd ridden him for three years and known him for six, and he never put a foot wrong. There was no harder rider than Bob Harrison, and many a time I'd seen him full gallop where few horses would have kept their feet. If that horse fell I'll bet it wouldn't have been his fault – some riders would pull any horse down.

How Bill finished up I don't know, but I hope he ended his days in a good home. For all Hanmer Atkinson's mishap he was a great horse and deserved the best.

# 8. The horse dealer

WITH six months to fill in before the next mustering season started I had to do something to earn a crust. I decided that a bit of horse dealing might be as good a way as any. I'd got to know old Joe Bunting of Woodend and he seemed to make a lot of money out of the game. I was to find out later that Joe had a great eye for a horse and that it was seldom that he made a poor buy. I've seen him pick up a scarecrow of a thing – one that I'd never have looked at – and in two months time he'd have it looking like a show horse. Anyway I reckoned that what he could do I could do too. Or so I thought!

The upshot was that I became a regular attendant at Tattersalls horse sale for the next few months. Although I felt pretty important I've got a sort of a feeling that my presence was more than welcomed by Jack Matson and company. I could usually be relied on to buy anything that no-one else would touch. Unfortunately it seemed that I was no Joe Bunting when it came to picking a horse.

My first Tattersalls buy was a pretty classy-looking thoroughbred mare from a well-known racing stable. I remember her name well, also the seller's, but I'll keep that under my hat as she turned out to be a plaster and my remarks could bring repercussions. Anyway with a very pleasant smile Jack duly knocked her down to me, and I climbed aboard and set out for Ohoka, 16 miles distant. I had been told that she was fast – very fast – but all I can say is that our progress was slow – very slow. That mare was as ignorant as a pig to handle, especially in traffic, and it was a long, slow ride home.

However I decided that I'd soon do something about that so the next morning I saddled up, buckled on a pair of spurs and at the first sign of any nonsense I sunk these hooks right to the maker's name – but only once. That mare may have been ignorant in some respects but when it came to bucking she knew all the answers. She was a light, whippy sort of a thing and when she did go into action she could tie herself in knots that would have put a cat to shame. Needless to add we parted company in very short time. The worst of it was that we were out on the hard road at the time and I can assure you that there isn't much bounce in a metal road – it's worse than bitumen. I know from experience. Fortunately there was no-one about at the time so I picked myself up, scraped the shingle out of my hair, and taking off those bloody spurs I very, very carefully climbed aboard again. That was the one and only time I used spurs on that mare.

Thanks to a fair bit of guile I managed to quit her soon after that, and what's more I only lost two quid on the deal. I put that down to experience. I sold her to a mate of mine at St James – when you're in the horse dealing game sentiment is out – and he also had reason to regret the deal. That mare must have been doped to the eyebrows for soon after that she lost most of her hair and nearly died. I met that bloke just recently, after 30 years or more, and although I'd have liked to know just what happened to her I thought it might be wiser not to bring the matter up.

After that episode I decided I'd do my buying privately. I'd heard that a Rangiora drover named "Hoppy" Young had a horse for sale, so I went along to give it the once-over. It was a rather attractive bay gelding and looked to be right in every way, so I asked Hoppy if he'd take me for a trial run. He yoked him up in the gig and drove right through the middle of the town; he went beautifully. Feeling that I couldn't go wrong with a drover's horse, I closed the deal. I led him back to Ohoka behind my push-bike, then keen to try him out I took him straight round to the stable and clapped a saddle on him.

The next thing I knew that saddle went sky-high, but thinking that something must have gone wrong I picked it up and tried again. This time I was more careful but the result was the same. Well, I discovered that that horse had never had a saddle on him in his life. He was as quiet as a lamb in harness but he'd never been ridden – what's more he had no intention of ever being ridden. However there were plenty of neighbours round about who were willing to give advice, and between us and a few lengths of rope we managed to get him a bit civilised. But he was always liable to have a go, and I concluded that Tattersalls was the best place for him. It was always reckoned that Jack Matson could sell anything.

Well, I regret to say that old Jack missed out this time. Hoping to recover my loss over the last deal, I opened my mouth too wide this time. The buyers seemed a bit suspicious and Star was passed in. As I mentioned earlier Tattersalls ran from Cashel Street right through to

Hereford Street, and from the selling ring there was a lane-way, 70 or 80 yards in length, which gave sellers an opportunity to show their horses' paces! Johnny Duncan, a well-known rough rider, was with Matsons at the time and when Star's turn came he offered to ride him. However feeling that it might be an advantage knowing something of his tricks I decided against this and climbed on him myself. We did a round or two of the ring and Star never put a foot wrong, but someone asked me to take him for a canter down the lane. That's when things went wrong. On the way down he went beautifully, but when we wheeled to come back I felt his back arch and I knew trouble was brewing. Fortunately there were other horses tied along each side of the lane so I wheeled him in among them hoping that no one had picked that there was anything amiss. But there were no flies on those horse dealers and the bidding came to a dead stop.

I thought I'd had it, but just as I was leaving a well-known dealer – who, I imagine, hadn't seen my performance down the lane – came up and asked me what I'd take for my horse. Thinking that I could always come down I stuck to my price and to my surprise he seemed interested. He asked me if I'd give a trial in harness, and of course this was right up my alley. This bloke – I'll call him Mac – lived out Spreydon way on the outskirts of the city, and it was arranged that I'd take Star out there to a paddock of his that afternoon and meet him there next morning.

I spent that night at the old "Blood House" down High Street – both the United Service and the Clarendon were full up – and I was out at Spreydon bright and early next morning. Early enough to give Star a good work out before Mac turned up. He arrived in his car soon after, and as he kept his gig round at his home he suggested that I follow him round. Anyway he took off in his car, but when I urged Star into a canter he started to perform. I was dead scared Mac would see him in his rear mirror but luck was with me. Mac ran a riding school for girls and if he'd spotted that performance he'd have wiped me smartly. Well, Star did his stuff in the gig and to my surprise I got my full price. I decided that horse dealing wasn't a bad caper after all.

How Star fared in Mac's riding school I don't know, but if those girls wanted to learn to ride they'd certainly have had the opportunity with that horse.

I kept away from Tattersalls for a while after that, but that place was like a magnet and it wasn't very long before I was back again. This time I bought a little chestnut mare which I named Blondie. A sturdy cobby type, she was a nice sensible mare and would have suited Mac down to the ground. I was tempted to contact him but decided that it might be pushing my luck too far. Eventually someone made an offer for her and Blondie and I parted company. She was a bonny little mare and if I'd had any sense I'd have kept her.

I put in two seasons at Castle Hill while I had her, and one night I did a journey on her that I thought wasn't a bad effort. My mate at Castle Hill was Jimmy Ferguson, a member of a well-known local family, and on this occasion we decided to ride down to Springfield to attend a dog trial committee meeting. We left after the evening meal and as the meeting was scheduled to start at eight we didn't have much time to spare. We did the 18 miles to Springfield in under two hours – only to find that the damned meeting wasn't until the following evening. Anyway we had a few beers – quite a few – and then climbed on our horses and headed for home. We got back about midnight after riding 36 miles and drinking a skin-full of beer. Little Blondie did it on her ear. Jim was riding a racy sort of a gelding that was by Masterpiece, but he had nothing on my mare.

It was soon after this that I saw a horse that completely took my eye. A great big chestnut gelding that really looked the part, he was owned by a neighbour of ours. One morning I went across to make some enquiries. This man, Bill Moir of Inglewood, Flaxton, was well known in the trotting game and in his time had raced one or two good ones. It turned out that the big chestnut, which was by a leading trotting sire, Rey-de'Oro, had been put into training but although showing a fair turn of speed had proved to be too difficult to handle. He had been given up as a bad job. He'd been turned out for about three years when I saw him and he looked a picture. Anyway, dirty or not I wanted him. Bill Moir was pleased to get rid of him and I got him cheap – or so I thought.

Well that horse, Ginger I called him, was just plain dirty. However a young fellow named Archie Smith was working around Ohoka at the time and as he was a bit keen on playing about with an unruly horse I enlisted his help. We had a lot of fun, as well as a few thrills and spills, and between us we managed to civilise him a bit. The biggest problem was that he stood a full 17 hands, and if he performed at all I couldn't get on him. I'd have to strap up a front leg and if Archie wasn't there it was a bit awkward getting it undone after I was in the saddle. Fortunately he couldn't buck much; he looked pretty ferocious, but he simply pig-rooted straight ahead and had no tricks. I was never any good in the pig-skin, but I could stick on Ginger alright – usually.

However down that way – among the cow cockies – a bucking horse was something of a rarity and those Ohoka blokes thought I was a real dinkum Queensland Harry. We lived only a couple of hundred yards from the township, and sometimes I'd turn Ginger up that way and stir him up a bit. He'd put on a bit of a turn. There would be women and kids peering round buildings with their eyes sticking out like dogs – I mean billiard balls – and me thinking I was really somebody. There was one occasion though when Ginger dumped me right in front of the store; there were half a dozen blokes about too. That took the smile off my face.

The real fun started when we tried him out in harness. I had a long shafter jogging cart, just the thing for a young horse, but all the same those shafts were none too long for Ginger. A 17 hands horse can reach a fair way with his hind legs. An iron-jawed brute of a thing, Ginger used to get away on me at times and when I did get him pulled up he'd put the boot in – and he *could* kick too. On one occasion he kicked both traces clean off the hooks on the swingle-tree and we did half a mile with only my grip on the reins holding him in the shafts. On another occasion he bolted when we were coming through Rangiora and, as he sometimes did, he struck the pacing gait. I reckon Indianapolis wouldn't have blown smoke up his behind that day. With the motorist blissfully unaware Ginger pounded along behind the car with his nose almost touching the rear window until, all of a sudden, the driver must have spotted him in the rear mirror. That bloke got such a fright that he jammed the accelerator almost through the floor-boards and his old crate just about took to the air.

I've often wondered if it was that incident that first got the trotting wallahs interested in the mobile starting system now used on the racetrack.

At that time Archie's old man, Dickie Smith, had Woody Gully – a little grazing run over Glentui way. One Sunday we decided that just as a civiliser we'd drive Ginger over there – a round trip of about 40 miles. On the way over we kept him bowling along at a fair pace; this must have taken all the sting out of him for on the return journey he peaked and it took us all our time to get him home again. For the last few miles we both had to get out of the gig and walk and even then it took a fair bit of belting to keep him going. Yet every time we belted him he still found energy enough to put the boot in. But he was a pretty civil horse for the next day or so. However it was the same story – once he got fit again he was as dirty as ever.

One winter a local farmer offered to take him and work him in his six-horse team so I jumped at the chance. That really civilised him, but I think it was due more to lack of tucker than hard work for when I got him back he was just skin and bone. But even that didn't work; he was soon his old self again and I gave him up as a bad job. I finally sold him to a bloke that rather fancied himself with horses, but I regret to say that this was another deal that left me in the red.

His new owner also worked him in the team, and according to one report he was going like a lamb – nothing wrong with the horse. However one morning this chappie had to shift the set of six-horse blocks and trees from one implement to another. Having led his horses into position he commenced to yoke them up. He had Ginger in the middle at the back and, as was customary, he was the first one to be hooked up. However just as the chains clicked into the hooks on the hames something startled Ginger and he went hay-wire. There was one hell of a tangle, but he finally broke loose from the rest of the team and with the whole set of chains and trees trailing behind him he took off round the paddock and then out the gate and up the road. The would-be horseman finally rounded him up in someone's yard a mile away and then had to spend the rest of the morning hunting for chains, blocks and trees. He took Ginger down to Tattersalls after that, pleased to see the last of him at any price.

But my story of Ginger didn't end there. About 14 years later – after World War II – when I was a land valuer with the Valuation Department, I had occasion to do a job out Rolleston way. The cocky in question had a big chestnut gelding that reminded me very much of Ginger and feeling sure it was the same horse I made some inquiries. This chappie told me that he had only owned the horse about 12 months and that it was 12 years old. This puzzled me as by then I was quite sure it was none other than old Ginger. I remembered that Ginger had once got hung up in a wire fence and had cut himself rather badly under the front pastern, so with the owner's consent I caught him and had a look. Sure enough, there was the telltale scar. That horse was old Ginger alright. The catch as far as his new owner was concerned, was that his 12-year-old was well over 20. However he told me that he'd had no trouble with him at all – he'd found him to be a good, reliable horse. Apparently age had done the trick. Anyway I gave old Ginger a friendly pat before I let him go – for old times sake.

If nothing else that horse had a ton of fight in him and better handled he may have been a good 'un.

Although the dealing game hadn't worked out too well I reckoned that if old Joe Bunting could make a go of it I could too. So I counted up my dwindling assets and decided I'd have one more go. This time I bought a lovely little cream mare from Bill Russell, now of Lees Valley. She was genuine in every way and I was able to pass her on with every confidence. She went up to St James to the same chap that had bought the hairless wonder. Fortunately he was more than pleased with her. Anyway it wasn't my fault that the black mare had been filled up with dope.

Well, that ended my career as a horse dealer. I decided that I'd retire from the game while at the peak of my fame. However I can't say I made big money at it; if memory serves me right I finished up about £3 to the good – not counting expenses.

I'd had a lot of fun, but that doesn't keep the wolf from the door. Eventually I was forced to do a bit of honest work. We were, of course, then in the dead of the slump and with jobs as scarce as hens' teeth that wasn't so easy. However my old boss R. O. Dixon had arranged to take a gang of men on one of the unemployment schemes and he wanted someone to take charge of them. I was silly enough to take it on.

The job was picking stones on a block that he had alongside the Waimak gorge bridge. Needless to add this could hardly be classed as a highly desirable position – especially as I had recently been in business on my own account – but it was a job. What did upset me was that some chappie I'd known in better days was heard to remark that I was in the right job at last – all I needed was a suit with a few arrows stencilled on it.

But at least I was still with horses. We were using the old tip drays and the shafters were two of the old Woodstock team, Billy and Pryde. There were nine of us in the gang altogether – eight on the 10A scheme (I think it was) and myself. We had two drays, four men to each, and it was quite amusing to see the way those blokes would all quietly gravitate to the far side of the dray whenever I joined them. Me working on one side and the other four "working" on the other. Even so there'd be more stones going into the dray on my side than theirs. However they were only getting paid ten bob a week, while I was getting the princely sum of £1. We camped in a mill whare, with a galley, and we had a cook – so I suppose we couldn't growl. The cook wasn't much good, but as he used to say, "How the hell can a man cook when he's got nothing to cook." We had to admit that he had a point there. But no-one was sorry when that job ended. The gang consisted of city chaps and, a bit down in the mouth, they weren't very good company. Even with Billy and Pryde to talk to I found that I was suffering from intellectual starvation.

Although I'd definitely finished with horse dealing I still went down to Joe Bunting's now and again, just to see what his latest purchases might be. It was while on one such visit that I got to know the Little brothers of Woodend. The Littles, father and sons, were probably the leading blacksmiths – particularly as farriers – in the whole of North Canterbury. When I first knew them – in the mid-'30s – the brothers Jack and George were running the business, but it had then been in the family for close on 70 years. James Little, the father, came out from England in 1863. He first established his smithy at Woodend in 1868, and he and his family were to carry on the business until Jack and George finally retired in 1937. They were then 67 and 65 respectively and had been plying their trade, first with their father and then on their own account, since they first left school. Old James was particularly versatile and won many awards for various innovations in the way of agricultural machinery – perhaps the best known being a bean hoe which he invented about the turn of the century. The Little light harrows were also in keen demand in the days of the horses.

But it was as shoeing smiths that the Littles were best known. They specialised in shoeing trotting horses, particularly those that needed some special fitting. It was common to see one of them out in the road in front of the smithy studying a horse as it was paraded up and down and then working out just what was needed to rectify some peculiarity in its gait. He would then set to and fashion shoes that would overcome the defect. Both Jack and George were masters of this particular art, and needless to add their services were in great demand among racing men. The Littles, of course, were operating in the days of the teams, and the shoeing of draughts was also big business. In those days there was a smithy in practically every township, and a good farrier was a person of note.

A third brother was also to spend practically his whole life in Woodend. This was Harley Little, who ran a chaffcutter in that district for just on 60 years. Harley actually lived in the same house – in the middle of the township – for over 80 years, and he was just as well known as Jack and George.

Two other people of that locality who were widely known in horsey

*Above:* The size of the task dictated the size of the team. Eleven horses shifting a portable engine to a sawmill up the Puhi River, Kaikoura. *M. Pearson*

*Below:* "The most famous horse ever to go up the Waimak. He and his rider earned the admiration of all back country men who knew him." The Rev. Alex McNeur on Fiddler (Mick). *Mrs L. Robertshaw*

*Above:* Burnfoot Farewell, the last Clydesdale stallion to be used on St James. *J. Stevenson*

*Below:* A most unusual sight. Four white pack-horses about to set off for the autumn muster at Mesopotamia Station in 1962. *F. Hampton*

circles were Fred McQuillan of Woodend and George Power of Flaxton. Fred, who always drove a classy gig horse, loved making a deal and there was always something new to look at in his stable yard. George did contract agricultural work with the horses for many years; he stuck to that game until well into the 1960s and must have been one of the very last to drive a horse team on contract work in this part of Canterbury. I remember him in particular for a fine big hack that he rode for many years. He was by Long Range – a descendant of the famous Carbine – and although a bit unruly he was a grand saddle horse and not a bad jumper. George, always full of confidence with horses, once tried my dirty big chestnut, Ginger, in a binder, but that definitely wasn't a success. Apparently George thought my talk of Ginger was all bull, but it didn't take him long to change his mind – it was his own binder he was driving.

There came a day when I pulled into Joe Bunting's, just for a yarn, and there in his yard was a little bay gelding that completely filled my eye. It was a beauty and sure enough he was on the market. I didn't haggle and before Joe had time to think twice we'd made a deal.

That horse was little Jimmy – the very horse that, after all these years, finally inspired me to write this story.

Without a doubt that was the best horse it has ever been my pleasure to ride. I kept Jimmy for 12 years, and although it is now 30 years since he ended his days that little horse still has a special place in my memories. As I stated earlier he was by a well-known thoroughbred sire from a pony mare and he combined all the best of those two breeds – speed, stamina, strength, ability and sheer guts. He was a beautifully coupled, showy little horse without an ounce of vice in his whole make-up. I've made him sound good – and he was good. To me he was the best ever.

Yet he had his faults. Joe Bunting had got him from a local boy who used to join up with apprentice jockeys who regularly took their horses out to the Woodend Beach for training sessions. He used to pit Jimmy against these horses, with the result that he became almost unmanagable in company; when Joe got him he was as mad as a maggot, and as Joe's riding days were well past nothing had been done to straighten him up. Indeed I was told by some of the locals to leave him alone – they reckoned he'd never come right.

But there are none so deaf as those who don't want to hear and I bought that horse. It proved to be far and away the best buy I ever made.

# 9. Jimmy – the greatest

I PUT in another season at Castle Hill after I purchased Jimmy and that gave me ample opportunity to try him out. His first test was fording the Waimak at Woodstock. Although it was comparatively low it is all in one stream at that crossing and carries a fair bit of water. As far as I knew this was the first time he'd been tried out in a sizeable stream – or any river for that matter – and I wondered how he'd shape. The water was almost up to the saddle flaps, with a fair current, but he breasted up to it well.

But it was a different matter on the hill. He'd gone clean mad with his previous owner and took a bit of managing. Having never before been on the hill, he scared hell out of me for a while. He had no idea of sidling, and when coming downhill he was unable to hold himself; he often finished up at a hand gallop. However he never once fell with me and I gradually gained confidence in him.

Indeed once he became accustomed to that sort of country he was a cracker – there wasn't a horse about that could hold a candle to him galloping in rough going. Not that I was a bold rider – I simply couldn't hold him when he got his blood up. As I said he never once fell with me but on one occasion, when galloping after cattle, both front feet landed in a little water course, hidden by tussock, and he went down on his knees; we must have gone a dozen paces before he regained his feet. It was a wonderful effort. All that enabled me to stay in the saddle was the speed at which he travelled while on his knees. Once he got accustomed to that sort of country I'd have backed him against anyone in rough going.

Castle Hill was then running quite a few cattle and they were by far the wildest and most difficult station cattle to handle that I've encountered. They'd been driven mad with excessive dogging and bad handling, and as the country on which they were running was interspersed with bush there were quite a few clean-skins among them. It was almost impossible to get a clean muster and the boss, Bob

Blackley, finally decided to quit all that we could get in. The big job, of course, was to get them yarded, but after a lot of hard riding we got a reasonable muster and a buyer came up to have a look at them. The buyer of those cattle was Eugene McDermott, well known in trotting circles. Not long after this, he collapsed and died while driving Colonel Gratton in the Trotting Cup at Addington. He was leading the field into the straight at the time and looked a certain winner.

The arrangement was that delivery was to be taken on the road at the homestead, the count being taken as they came out of the yards. They got the count alright – but not the cattle, for once on the open road the mob broke in all directions, over fences and everywhere. There were four of us on the job – Jim Ferguson, myself and two drovers – and between us we managed to keep them going in the right direction, although until we got them beyond the paddocks and out into open country they were flattening fences in their stride. One of the drovers, a young fellow named Alex McConnell, had a mare which by all accounts was a cracker. Alex boasted that there was nothing in the country that could show her the way to go in rough going. He had the opportunity to show her worth that day, but he had to pipe down when he got up against Jimmy. His mare couldn't foot it with him.

I can't remember the number of cattle in the mob or what were finally delivered, but I'd think that anything up to 20 percent got away. We never really got control of them until they were past the Springs yards on Brooksdale – about four miles from the Castle Hill homestead – and by then half of them were peaking and planting in the scrub. Jim and I had been told to assist the drovers as far as the foot of the Pass; odd ones were knocking up all along the way. On one occasion an old cow planted in a clump of gorse and as she wouldn't budge Alex got off his horse and put a match to it. The gorse, as dry as tinder, burst into flames and and the next moment the cow charged out – heading straight for Alex. He made a flying leap for his hack, landing spread-eagled across the saddle as the mare took off. The cow missed them by a whisker. I'll concede that that mare was pretty slippery that time.

Jim Ferguson was riding The Rat that day and he was no slug either. A racy sort of a gelding by Masterpiece, he'd been brought to Castle Hill by an old school-mate of mine, Alan Hawkin. Although I don't know where Alan got him, we understood that he'd once been raced. He can't have made the grade, although alongside the station hacks he looked a real Phar Lap. On good level going my horse never had a show with him. I've never seen a horse that flattened like The Rat when full gallop. His belly almost seemed to touch the ground. Alan must have sold him to Bob Blackley when he moved on; Bob often talked of putting him into training again. I don't know where he finished up, but at one time Gordon McCready, who was then with Pyne and Co. in Christchurch, was interested in him as a polo pony. He was only a little horse, pretty level headed, and with his speed might have made a good 'un.

Bob Blackley, the boss, rode a tremendous big bay gelding called Digger. A trotting-bred horse; he'd have passed for a stallion. But for all his impressive appearance, he was about as useless a hack as you'd get. A straight-out natural pacer, he was all action; with his legs, mane and tail flying in all directions he always reminded me of a threshing machine going full bore. But for all his action you could kick your hat as fast, and on the hill he gave the impression that he was going to turn a somersault at any moment. When after cattle Bob would be dragging the chain miles behind. Whenever Digger took off Bob would lose his hat, and he seemed to spend most of the time getting off to retrieve it. However he thought the world of Digger. I suppose that's all that mattered.

It was while at Castle Hill with Jim Ferguson that I once again became interested in the racing game. The Fergusons had always been keen trotting men and it was not long after I first went up that way that Jim started racing his own horses. When I was with him he had a bay gelding, Sports Review, which recorded some good wins in saddle races. These were few and far between – whether by design or otherwise I don't know – but when he did pull off a win he usually paid a hatful. Unfortunately I always backed him on his "off" days; I never once had a penny on him when he did come home.

His present horses stem from a Gratton Loyal mare called Bounty. She left Lady Bligh, which was probably the best Jim had. She recorded some good wins and was a real money spinner. She left a good mare, Prim, who in turn left another good one which he named Proper. After winning several races Proper was recently sold to American interests. Jim's latest hope is another young-'un out of Prim. Here's hoping she turns out a cracker.

Although Jim really knows his stuff in the trotting game it seemed that he'd had little to do with the gallopers, and one morning when we were riding out I bet him that I could name more thoroughbred sires than he could trotters. Of course Jim could reel off trotting sires almost without thinking, and knowing that I was not a real racing enthusiast he jumped at it. Taking turn about we got going. Needless to say there was no stopping Jim, whereas I knew only a scant half dozen or so. However without batting an eye I matched him horse for horse. The story, of course, is that all those that he named were genuine whereas mine weren't – I was simply making up names. Anyway I must have reeled off a couple of dozen when I came up with Johnny Larkin and Jim suddenly tumbled to my game. But Jim was starting to think, too, and if I'd been able to come out with another two or three I reckon I'd have had him stumped.

There was another young fellow shepherding in that locality who was also destined to make a name for himself in the racing game. This was Oscar Watson, who was at Snowdon – in the mouth of the Rakaia Gorge – when I first went to Lake Coleridge. Oscar then had a pretty classy gig horse that was by Logan Pointer. It seemed that even as a boy he was wrapped up in trotting horses.

Anyway sometime in the early '30s he took over his father's farm at Annat, and with his enthusiasm not to be denied he managed to scrape up enough money to buy a well-bred trotting mare. Even then Oscar had ideas of going in for breeding, but in order to get the cash he had to convince his bank manager that he was badly in need of a gig horse for his work on the farm. This mare, Peggywood, cost Oscar the princely sum of £25, but it was to prove one of the best buys in the history of the trotting game. That mare was the foundation of Oscar Watson's Rosewood Stud, one of the greatest breeding establishments in this country.

Peggywood, who was by Parkwood from Mary Wood, was mated with the Jack Potts horse Gamble on two occasions – 1940 and 1941 – and this produced Lady Ngaio and Rosewood. Lady Ngaio, who had the misfortune to lose her life when only six years of age, established herself as one of the fastest mares in the country and had an impressive list of wins to her credit. But it was Rosewood who really put the stud on the map. This mare was never raced; she was kept solely for breeding, and of her 15 progeny eight were highly successful on the racetrack. Of the others several who were never raced also turned out to be outstanding brood mares. One of them, named Torlesse, was the dam of the famous Good Chase.

Peggywood lost her life in extraordinary circumstances. There was an old "hole-in-the-ground" dunny in the yard where she often ran, and finding the door open one day she stepped inside; the rotting floorboards gave way and she fell head first into the hole and drowned. At the time Rosewood was only six weeks old and had to be reared on the bottle.

I often visited the Watsons and one of Rosewood's progeny that took my eye in particular was a dark chestnut colt called Red Chase. This horse broke down while still in his prime, but he recorded some impressive wins and became noted for his great finishing bursts. There's a story about that horse that rather tickles my fancy. On this occasion Oscar had taken him to Reefton and during the morning he gave him a round or two of the track for a warm-up. Oscar, who was a big well-fed sort of a chap, was driving him himself. He invariably galloped Red Chase while doing this, and as he passed a group of spectators one of them was heard to remark, "Look at that big fat bloke galloping that horse. What the hell is he thinking of – he'll never do any good with him." Red Chase duly won that race – the Reefton Cup – in what was then an Australasian record for that distance.

But it was with Good Chase that Oscar Watson really made his mark. This big black horse, which is by Light Brigade from the Rosewood mare Torlesse, has a remarkable record both on the track and as a sire. He raced in this country until five years of age and was the top two-year-old and three-year-old of his time. He was then leased to United States interests for five years and his winnings in that country – including several successes at the famous Yonkers Raceway – amounted to just on $170,000. It was stipulated that he was not to be used for stud purposes while in the States, but since coming back to New Zealand he has been kept solely for that purpose.

Largely due to Good Chase the Rosewood stud has met with phenomenal success at the yearling sales. On at least two occasions their entries have topped the sale; in 1968 a colt by Local Light from Torlesse realised the then record price of $9500. And at only 12 years of age Good Chase is still going strong. I'd say that this is a success story if ever there was one.

Oscar Watson died well before his time but fortunately his two sons, John and Denis, have inherited their father's flair with horses. The Rosewood stud is still making the headlines in the racing world.

If I'd worked with Jim Ferguson much longer I might have become a real authority on the game – and perhaps have made a little money out of it. However I do remember that it was about that time that the famous Harold Logan was at his peak. He won the New Zealand Trotting Cup in 1931 and 1932 and was the idol of the Addington crowds. I remember that mighty little horse in particular as he was bred at Springfield. He was born behind the pub when it was owned by his breeder Jack Coffey. Red Shadow won in 1933 and then came the mighty Indianapolis with his record of three wins on end – 1934, 1935 and 1936.

While at Castle Hill I used to give a hand with the mustering at both Ben More and Brooksdale, two adjoining stations. Ben More was then owned by F. M. Robinson, who became well known as "One Chop" Robinson. How that came about I wouldn't know, as during his time at Ben More it was just about the best tucker place I've been on. When mustering Ben More itself (the mountain) we spent a night out at the Acheron Hut, and on one occasion the cook sent out a big apricot pie. Always a glutton, I had a fair go at it that night; apparently this had not gone unnoticed. A day or two before the next muster Robbie told the cook he wanted an extra big apricot pie sent out to the hut as there was a young fellow coming down from Castle Hill who liked three helpings. Whatever faults Robbie may have had he certainly wasn't mean with the tucker.

Another memory of Ben More is of Robbie's hack. It was a beautiful, big, dappled grey gelding, very similar in appearance to Davie Pringle's grey at Mt Peel, and it also was a wonderful walker. This is one of the essentials in a shepherd's hack – as well as the fact that it should be easy to sit on. A chap called Alex Pow came to Castle Hill about that time; his hack, a great big leggy mare, could outwalk anything on the place – but man, she was just terrible to sit on. Although she could certainly cover the country she felt as dead as a door nail. In comparison my hack Jimmy, although not nearly as good a walker, was all spring and life; no matter how long that day I never tired of riding him. At lambing time he was a cracker. In hill country this is simply a case of riding all day and every day, for weeks on end, with little else to do. On some horses it can become very boring and tiring. This was where Jimmy shone; at the end of the lambing season he would be swinging along just as free as ever and was always lovely to sit on.

He had the one bad fault – he was mad-headed in company. At times when coming home in company he became so unmanageable that I found it easier to get off and lead him. But all the same he'd do me.

There was one occasion, however, when he came within an air of ending my days. At the time I was full gallop across a tussock flat trying to head off some cattle. In the excitement I'd clean forgotten that the telephone line crossed straight ahead of me. Like most back country lines it was a single wire; with the span between posts being several chains the sag in the middle was usually only a few feet off the ground – not much higher than a horse's wither. Anyway I suddenly spotted it, only a chain or so ahead, and with not a show in hell of pulling Jimmy up I ducked low in the saddle and hoped for the best. That wire actually scraped the back of my head. If it had been 6 inches lower I'd have been decapitated as clean as a whistle.

Les Milliken, who managed Brooksdale estate for the family, also had a string of hacks that would compare with any. When mustering their back country – out beyond Porters Pass – we camped at the Springs yard hut, and it was the practice to cart out all the camp gear in an old four-wheeled buggy drawn by two horses. For many years these two were Creamy and Dolly Grey. Creamy, a big solid type of a hack, was a beautiful horse in every way; Dolly was a sturdy, cobby sort, also a lovely mare. That pair really took the eye when in the buggy. Les seemed to name his horses according to their colour and he had another fine big hack that he called Brownie. They were three good horses.

Brooksdale also ran two six-horse teams and while I was there who should take on the teamsters job but old Jack Halliday. Jack, a happy-go-lucky Irishman, had driven the No. 3 team at Charlie Leach's before I took it over. With his ready wit and, more often than not, a lilting Irish song on his lips, he was a great old bloke and wonderful company. I never saw him again after he left Brooksdale.

What must surely have been the most famous horse ever to go up the Waimak was a big bay gelding named Fiddler, which was owned for some years by the Reverend Alex McNeur. McNeur, who was then stationed at Annat a few miles below Springfield, had as his parish what was known as the Central Ridges. This took in the heart of the back country right from Lake Wanaka to Takaka in Nelson, as well as the West Coast back country from Top House to Jacksons Bay and the Haast Pass. The round trip, during which he visited all the remote corners in that vast area, was done on horseback and must have totalled 800 miles or more. I believe that he tried to do the round twice a year. An incredible undertaking.

The Rev McNeur was five years on that job – from 1919 until 1924. Fiddler – or Mick as he called him – was his hack for the whole of that time and his reputation was such that I feel his story should be told in detail.

Fiddler was by a well-known trotting sire, Son of a Gun, out of a mare named Tahu. From the start he showed fire and determination and it was felt that he had possibilities on the racetrack; however he proved to be too difficult to handle and was finally given up as a bad job. He was eventually sold at an Oxford horse sale, but his purchaser discovered that he was a semi-outlaw and refused to complete the deal. A Springfield man, Andy Russell, who was a great hand with a horse, undertook to take him and in his hands he became a different horse – good in harness and a grand hack.

It was soon after this that McNeur was posted to Annat. One of his first jobs was to find a hack that had the qualities which the rounds of his unusual parish demanded. He contacted Andy Russell and it looked as though Fiddler would fit the bill. He stood a full 17 hands and was solidly-built and there was no doubting his stamina and courage. His one drawback was his contrary nature, but as McNeur was a good horseman he was not deterred and a deal was made.

Although the horse had been registered as Fiddler, McNeur renamed him Mick and it was by this name that he became best known. During the next five years Mick proved himself to be a remarkable horse. He and his rider earned the admiration of all back country men who knew him. For all the thousands of miles he covered, under the most testing of conditions, Mick never lost his fighting spirit and if given a day or two's spell would still have a go at unseating his rider. Fortunately McNeur was a good rider and it was this as much as anything that made his name. If there is one thing in particular that earns the respect of back country men it is the ability to handle and master an unruly horse. McNeur was to prove himself on many occasions.

Alex McNeur himself used to tell one story of his horse and himself that amply illustrates the high regard in which they were held. On this occasion he was riding along a back country road when he was overtaken by a man in a gig. The man, who was unknown to McNeur, pulled up and jumping out walked across and patted Mick with genuine affection. He then turned to McNeur and offering his hand remarked, "There are two things I've wanted to do for a long time. One was to pat your horse and the other was to shake hands with you."

When he finally left Annat McNeur handed Mick back to Andy Russell. He duly sold him to Bill Nimmo, a local musterer who was to ride him for the next five or six years. He also tells a story that is typical of this remarkable horse. At the time Bill, who lived in Springfield, was working at Mt Torlesse Station, some six miles west of the township. He regularly drove Fiddler backwards and forwards. Annat, Fiddler's old home, was a few miles to the east and Bill tells me that as he came out of his gateway onto the road he could do nothing about it if Fiddler decided to turn towards Annat. Although an experienced and capable horseman, Bill had no option but to let him go. After half a mile or so he was able to turn him, and Fiddler would then head away up to Mt Torlesse without any further bother.

Bill eventually pensioned him off at Flock Hill, one of the stations on which he had been working, and it was there that the old fellow ended his days. Game to the very last, he actually dropped dead while being ridden by a visitor to the station.

Alex McNeur himself died only a few years ago, but it will be a long while before he and his horse Mick will be forgotten in the back country. One of his family's prized possessions is a framed scroll which was presented to their father by the Presbyterian Church in appreciation of the wonderful job he did while minister of that vast area. Few men of his calling could have done it – few would have done it even if they could.

At Castle Hill there was always a break in the work after shearing was finished, and during my last season there I took this opportunity to ride through to Mt White to see if I could book a mustering beat for the following season. In order to see some new country I planned to ride right through to Eskhead Station and from there through the Virginia country and out to Oxford via Lees Valley. The trip, right home to Ohoka, would involve a ride of just on a hundred miles and would be a good civiliser for Jimmy. That horse needed all the work I could give him.

From Castle Hill I simply followed the road to Flock Hill and from there swung across country, round the southern slopes of Mt Bernard and out to Craigieburn Station on the West Coast line. My track from there took me out onto the Grassmere ewe country and down into the Waimak, which I crossed just above its junction with the Poulter. Fording the latter river at the old waggon track, half a mile below the bridge – which spans a narrow nick of gorge a full hundred feet above the stream – I had only to follow the road to the Mt White homestead.

The ride from the Craigieburn homestead to the Waimak – a

distance of between two and three miles – proved to be of particular interest to me as an illustration of the remarkable memory and sense of direction which horses seem to have. I did that ride in January 1935 and Jimmy was never over that country again until April of the following year – 15 months later. On that occasion I was riding back to Craigieburn at night and in the dark I went astray. I finally decided to give Jimmy his head and without hesitation he wheeled round and went back to where I'd gone wrong. Although there was no track which he could follow he took me right to the Craigieburn homestead. I have often seen evidence of this "sense" in a horse, but I thought this case quite the best I've known. That was another mark in Jimmy's favour.

The following day, with the Eskhead homestead my destination, I simply followed up the Esk River to Anderson's camp, crossed the Dampier Range at the Wollies and dropped down to Deep Creek – Eskhead's main mustering camp. From there I followed a well-defined pack-track down the south branch of the Hurunui to the Eskhead homestead. All told a ride of just on 40 miles.

That visit to Eskhead is also worthy of some comment. The homestead itself, a great cob building of six or seven rooms, must be one of the few station homes built of cob that is structurally still in its original state. It is now about five years since I saw it last, but as far as I am aware the only alteration is that the thatched roof has been overlaid with corrugated iron and some flooring has been renewed. Inside, the "ceilings" are simply the original thatch and the low doorways and tiny windows are the same as the day it was built. The cob walls, about 2 ft thick, are still as sound as as bell; it looks as if it could stand for another century.

Another memory of that old house is that as I entered I noticed a neat row of perhaps six pairs of slippers. These belonged to the musterers and station hands. Every time they entered the house for their meals, they discarded their boots, lined them up outside the door and donned their slippers. They did this without fail – and without demur. Angus Trumper was then managing Eskhead, but it was his wife, a wonderful woman, who was responsible for the co-operation of the men. I've never forgotten that.

The next morning I headed down the Hurunui and turning up the Seaward, which branches off to the sou-west, crossed over into the Waitohi at Virginia. At the head of the Waitohi I crossed a low saddle and dropping down into the Okuku, just below The Den, struck out through the Glenard country and out to Glenburn. I spent that night at Glenburn with the Harry Scotts – a great port of call – and it was there that I first saw a woman "roll her own". She had a one-pound jar of Greys tobacco on the mantelpiece and, as naturally as the day, she'd get the "makings" and roll cigarettes as well as any man. Another of those odd memories that stick in one's mind.

On the third day I had only to ride down the valley bottom and follow the road out through the gorge to Oxford, some 20 miles from home. Lees Valley lies between a belt of bushclad foothills and the high Puketeraki (Pakety) Range. With a sunny day and a free moving hack I found it a lovely ride. During World War I the valley had been taken over by the Government and cut up for soldier settlement, but with the blocks too small for that class of country the original settlers had a grim struggle. At that time the only habitation in the whole length of the valley was two musterers' huts. As the eight settlers were then entirely dependent on horse-drawn transport the setting up of their holdings was a formidable undertaking. The top end of the valley is just on 30 miles from Oxford, and for eight or ten miles the road – which had only just been completed – wound its way round the narrow confines of the Ashley Gorge. I take my hat off to those Lees Valley settlers.

With considerable amalgamation of the original blocks Lees Valley has today really come into its own. Yet for all the improvements and amenities of modern times it is still one of the few localities where the surplus stock are driven out "on the hoof". The stock route is over the bushclad Richardson Range – the Blow Hard track. The horse still plays a vital part in the running of "The Valley".

# 10. Gandhi – the worst

MY plans to go to Mt White had eventuated, but as I had a few months to fill in before the mustering season started I joined up with my old mate Jim Binnie and we put in a few weeks hunting wild cattle and breaking in horses at Lake Coleridge. Two jobs with plenty of excitement.

Although the "Lake" was not then running cattle, a few from earlier years had found their way up the Avoca and Harper Rivers away at the far end of the run. These had eventually gone wild and over the years had increased to a hundred or more; they were clean-skins, all of them, and of course half of them were bulls of all ages. The arrangement was that Jim and I were to get out what we could – alive – and then return and shoot the rest. We finished up by getting between 60 and 70 and shot 30-odd.

As I have previously written of this experience at some length I will now touch on it only briefly and try to confine myself to the part played by the horses. It was on this job that Jimmy really proved himself. He was too rattle-headed to be a good stock horse, but on this job – where speed, reliability and guts were the real essentials – he was a cracker.

There are some good stretches of flat up the Avoca and as it was there that the majority of the cattle seemed to congregate it seemed that our chances of handling them were pretty good. How little we knew. These flats were bordered by areas of bush, but we reckoned that when the opportunity arose we could sneak between them and the bush and with one good rattle wheel them away from this refuge and down river for a few miles. It sounded good strategy but it just didn't work. At our first attempt we swung them down river as we had planned, but they finally wheeled back towards the bush. Once they had their compass set all hell wouldn't have stopped them. I was foolish enough to try, but it was a forlorn hope and I was lucky that it didn't end in a disaster.

The flat was criss-crossed with little terraces and dry water-ways, most of which were hidden by matagouri and heavy tussock; how Jimmy kept on his feet I wouldn't know. To make matters worse some of those bulls were pretty aggressive and had Jimmy fallen anything could have happened. When a chap's blood is up he doesn't think of these things – until afterwards. Suffice it to say we only tried that caper once.

We eventually found that our best policy was to leave the job entirely to the dogs until we got the cattle well away from their stamping ground. Each morning we'd sort out a little bunch – seven or eight – and keeping well out of the way let the dogs work them. When they became a bit tractable we'd take them down the river for several miles. We'd play around with them until dark and then leaving them in some sheltered spot we'd head back to our camp. Next morning we'd be back at them by daylight and give them another handling lesson before we headed away up river for another lot.

Jim's hack was a sturdy sort of a grey cob. The Binnie boys had used him a lot, snigging strings of horses from sale to sale when they were with Linc Wright. He was worth his weight in gold on that job, but on wild cattle he was too slow – in rough going he was hopeless. If a beast turned frosty Jim didn't dare go within chains of it. However Jim had the bobtailed heeling dog that they used on the sale horses and he was a real asset on those cattle. He had no ability when it came to working stock, but if a single beast happened to break from the mob Jim would put Bob on it and that beast would almost split the mob in half getting back into the safety of its numbers. He was the severest heeling dog I've ever seen. When we finally took the mob into the station they were pretty civil cattle – thanks largely to Bob.

Bill Parkinson, a well-known cattle dealer from Kaiapoi, purchased those cattle and Jim and I had to deliver them to Springfield. We went out via Porters Pass and knowing that a Castle Hill bullock had been lost in the Kowhai River bed when we took their cattle out a couple of years ago, I took it upon myself to pick it up as we passed – if I could. I'd seen it at odd times when passing, and knowing just where it was to be found we managed to join it up with our mob. Bill Parkie and his drover met us at Springfield and he agreed to take it at what seemed a reasonable price. Anyway I thought I'd done Bob Blackley a good turn. To my dismay I found that Bob had recently sold it to Ben More Robinson and I had the unenviable job of telling him what I'd done. Actually we'd had done him a good turn as I heard later that they'd had several attempts at getting it in, without success, and it had become a hermit. But the way that old bloke Robbie performed I thought he was going to have me up for cattle stealing.

What with horse stealing as a boy, and now cattle rustling, it looked as if I was a pretty undesirable character.

It's worth noting that although these cattle were all genuine clean-skins – until the last few days they had never seen a fence – Bill Parkie's drover Les Chaney took them right to Kaiapoi single-handed. What's more he had to take them over the Waimak Gorge bridge.

At this time "The Lake" had a fair string of unbroken horses. As some of them were getting on in years Jack Murchison, the boss, decided it was time something was done with them. Jim was a good man with a horse so we undertook to break in a few of them. They were then running on a block known as the Peninsula – about halfway up the northern side of the lake – and as none of them had ever been handled there was some hard galloping before we finally yarded them. One racy sort of mare was particularly determined; although I just about galloped Jimmy into the ground she was too many for me. I could head her off but she just wouldn't turn.

They were an assorted lot, ranging from the real pack-horse type to light hacks, but there were some nice horses among them. The pick of them, by far, was a beautiful bright bay with a white blaze and four white stockings. A big strong horse – almost the pack-horse type – it was said to be about eight years old and it was obvious that it would take some handling. Jack Green, the head shepherd, claimed it for himself, but its age was against him and it proved to be so determined and full of fight that we finally gave it up as a bad job. That horse just fought and fought and never gave in. In his day Jack Green was a pretty good rider, but he'd injured his back pulling on the lazy-stick a year or two ago and was past doing battle with a wild horse.

A year or two later Bill Russell, a professional horse-breaker – and a pretty good one too – had another go at that horse. But he also gave it up as a bad job. It was finally shot for dog tucker. I couldn't help thinking of that old saying about the white stockings: "One buy, two try, three doubt, four better without". But what a pity he hadn't been taken in hand when young. He was a beautiful horse and would have made a hack anyone would have been proud to own.

With Jim as the "master" the other three gave little trouble. One, a big black gelding, was intended for the pack-team and he also was a fine stamp of a horse. Of the other two, one was a sturdy cobby bay without a single speck of white on him and the other was another black. All three were nice, sensible horses and should have been a valuable addition to the station's string.

Unfortunately it didn't work out that way. When we handed them over they were going well and only needed regular work to finish them off. We turned them out in a spare paddock the night before we left and it seems that none of them were touched again for about two years. Needless to add they took some handling and the black hack, in particular, turned really dirty. He used to throw himself when being ridden, and on one occasion this resulted in his rider getting his leg broken rather badly. Black Jack, as he was known after that, was also given up as a bad job. The bay cob, although a bit of a problem for a white, was finally purchased by Jim Ferguson. He turned out to be a lovely level-headed horse – an ideal shepherd's hack. Jim had him for quite a few years.

So much for breaking in horses. If they don't turn out right the breakers got the blame. But it's not always their fault – the man who takes them over has got to do his stuff too.

I might add that Bill Russell, whom I'd known for years, spent a year or two in the North Island. Among other things he had a go at the wild horses up Taupo way. The year following that he put in a season in the flying gang at Hanmer and it was arranged that he and I would go back up there at the end of the season and have another go at them. Unfortunately Bill contracted a bout of rheumatic fever about that time and our scheme never eventuated. I was dead keen to try Jimmy out on those brumbies.

When we finished at Lake Coleridge Jim took off for Mesopotamia. Before heading into Mt White I headed down country for a few days. Jim was travelling to Meso by car so I took his hack to Ohoka, where it could be picked up by one of the Binnie boys. It's 60-odd miles from "The Lake" to Ohoka and I reckon that that ride was the worst I'd ever undertaken. That's a long way to snig a horse at the best of times, but that horse, Grey he was called, was the worst leader I've struck. In the end I decided to ride him and lead Jimmy, but if anything that was worse – that bloody Grey was a real slug and my heels were going all day. Unfortunately Jim had taken Bob – the heeling dog – with him. He'd have made that slug of a horse lead. I finally got home about midnight after 16 hours in the saddle. What with getting my arms nearly pulled out for the first half of the journey and thumping his ribs

*Right:* Jimmy, the horse that inspired Peter Newton to write this book. "He was a beautifully coupled, showy little horse without an ounce of vice in his whole make-up." *P. Newton*

*Below:* "It's good to see them coming into their own again." Mr Conner's four-horse wagon team several years ago at the Christchurch Show. *The Press*

*Above:* The mighty Indianapolis, winner of three successive New Zealand Trotting Cups in the 1930s. *G. Weigel*

*Below:* Cardigan Bay, the "million dollar" horse. *G. Weigel*

A good drover's horse would simply trail along behind the sheep, grazing as it went – a real asset. This mob is in the Waioeka Gorge, Bay of Plenty. *Alexander Turnbull Library*

The glory of the Clydesdale. James Wyllie's Bonnie Kincardine showing his paces in the 1930s. *J. G. Wyllie*

non-stop with my heels for the other half, I'd just about had it. But we got there.

That job at Mt White was the culmination of an early ambition. This was to muster on four places which for some reason or other had captured my imagination – Lake Coleridge, Mesopotamia, St James and Mt White. At the time it seemed ambition enough and I could perhaps add that of my many experiences nothing, before or since, has brought me more pleasure than that simple achievement – and I'll chuck in a trip round the world.

Mt White is another station that in earlier years built up a great reputation for its horses. As a boy I'd heard of their merits in the days of the horse-drawn trams, and they were also in great demand with the old carrying firms. Like the St James horses they were then mainly a quarter-draught; another similarity was that in colour they also were predominantly dark brown. You seldom saw a bad one of that colour.

In my time at Mt White they had cut down on horse breeding, although they still ran about 15 brood mares. These were a light draught but the station had been using a Clydesdale horse for some years and in type and colour they had changed considerably from the original horses – not for the better either.

However they still had an attractive line of mares; all of them were bright bay with white blaze and stockings. From a distance they looked as alike as peas in a pod. These mares were always run up the Poulter and as they were yarded only once a year – at weaning time – they were as wild as deer. At the sight of a rider they'd take off up river, and with their coats glinting in the sun and manes and tails streaming, they were a fascinating sight. The Poulter River, a branch of the Waimak, stretches as straight as a die for 20 miles, right to the forks where it disappears in bushclad ranges. The lower reaches are flanked by good flats, great country for horses, and these mares are always in great fettle. They were always on the alert for riders, and at times, when they had foals at foot, we've had to go right to the forks before we could get them headed off. Mustering and weaning those mares was one of the best jobs of the whole season.

At that time the station was using a Clydesdale stallion known as Hector. This horse was bred by Hector McIntosh of Kaiapoi – one of eight brothers, all of whom became well known in Clydesdale circles. Their father, Kenneth, who came out to this country in 1859, was also a prominent horseman; he imported two top Clydesdale stallions well before the turn of the century and his sons – particularly Hector and

Kenneth junior – bred Clydesdales for many years. One of the brothers, William, lost his life in World War I but the other seven all farmed in the Rangiora-Kaiapoi district for the whole of their lifetime. Few families are more widely known in this part of Canterbury.

The young stock – two and three-year-olds – were run on Lower Pakety, a beautiful sunny block straight across the Esk River from the homestead. Every now and then they were mustered in; colts had to be gelded and the sale draft cut out and handled. Jim Thompson, widely known as "The Bully" – I don't know why – was then managing Mt White. Although only self taught, he always castrated the colts himself. There were about half a dozen every year and although it's generally considered to be a ticklish job – requiring the services of a qualified vet. – none of the many that he did seemed to come to any harm. He wasn't very fussy either – he just threw them, roped them up, got cracking and got the job done.

Lacking the guts and spirit of the old "Mt Whiters", the young horses were lovely and easy to handle. Although they were probably a good proposition for the average farmer they never gained the reputation that was a feature of the original Mt White horses. One of the old Mt Whiters, a mare that was past breeding, was run with the young stock, mainly as a leader when mustering them into the yards. She was a nice cut of a mare, typical of the type – a dark brown quarter-draught – and for all her age showed twice the fire and determination of the others.

A thing that may account for this gradual loss of "guts" in the Mt White horses is the gradual change in the feed on this country. It seems that the Poulter valley, in particular, once carried a good sprinkling of bluegrass; this has now almost disappeared. This grass is famed as horsefeed (horses off the Kentucky bluegrass country are world renowned). I don't know bluegrass when I see it, but a visitor to the station – George Armstrong, one time of Seddon – once pointed it out to me with a remark that no other grass could compare with it for the raising of horses.

Jim Thompson had as his own hack a handsome big thoroughbred mare by Masterpiece. On one occasion he mated her with Paper Money. The foal, a colt called Peter, was reared on the station and as it turned out to be a pretty likely looking sort Thompson decided to put it into training. I can't remember whose stables it went to, but it just lacked the heart and guts – and the speed – and was eventually sent back to Mt White. I was sent out to the Cass railway station to pick it up and on the way in I gave it a try out on a good stretch of flat. I could have ridden a kids' trike faster – and carried a pig on the handlebars. Well, they bred from that horse for some years but it didn't work out. I blamed the horse, but old George Armstrong might have been right. He was a pretty shrewd, knowledgeable old bloke. Yet strangely enough that country did other stock well.

Although I don't think that the actual grazing was responsible it so happened that when I first went to Mt White they had the most hopeless string of station hacks I've ever struck. Jim Thompson's big thoroughbred Joan was alright and there was another aged mare called Lucy that was still a good hack. But the others were a sorry lot. There was a big raky chestnut named Tan, a dead and alive bay called Tom, another chestnut, Phar Lap, that even the dogs would have been loth to eat, and to cap them all, the dirtiest, most ignorant pig of a horse it has ever been my misfortune to ride.

This horse, Gandhi, was a real slug but it was also just chockfull of dirt. He couldn't buck for nuts; in fact if you were ready for him he wouldn't even try. His long suit was to wait until his rider was miles away – dreaming of my darling love of thee or something like that – and then when he reckoned the time was ripe, one heave would send his rider flying. As I said he couldn't buck his way out of a paper bag, but, oh boy, he had some victories. Few who rode him could boast that he'd never dumped them. Joe Casey might have been able to ride The Brown Colt, but Gandhi dumped him too.

Old Gandhi had two victories I'll never forget. On one occasion the head shepherd, Arthur Brown, was riding him when mustering and during the day one of his dogs cut up a bit rough. Arthur used to perform a bit when a dog went wrong – he's worse than me, if anything, and I've been reduced to tears often. This time he was fairly frothing at the mouth. Anyway old Gandhi thought this too good a chance to miss. With one mighty heave he sent Arthur flying. Such was Arthur's rage that he never stopped swearing, and when he landed both hands were clutching for stones to throw at the dog. I don't think he even knew he'd been thrown.

On the other occasion Murdo MacRae, Clive (Honey) Richards and I were camped at Riversdale, a Mt White out station only a few miles from the Cass railway station. A beautiful spot too – makes my mouth water to think of it. Anyway in those days station tucker wasn't the best so we decided to pool a couple of quid and get a few extras sent up from Springfield – a bit of bacon, a couple of dozen eggs and a few bottles of "medicine".

The goods duly arrived at Cass and Honey, who was then riding Gandhi, saddled up and went across to get them. An hour or so later he arrived back with a carton balanced on the pommel of the saddle in front of him and reported that everything was safe and sound. Just as he was handing the carton to Murdo, Gandhi sensed the opportunity and – wait for it – one mighty heave had Honey, carton and all, flying through the air. Needless to add everything breakable was broken. That damned old horse had never put a foot wrong, going or coming, but true to form he picked just that moment to do the dirty. Bang, we were back to mutton and spuds. We could have cut his bloody throat with a blunt pocket knife.

In mentioning "Honey" Richards it might be worth recording just how he came by that nickname. At the time he was cow-spanking at Mt White, only a lad, and a certain joke that he'd heard so tickled his fancy that he was always repeating it. Like Honey I've told this story before but I feel it bears repeating. It goes like this:

A young fellow who was spending a few days at a certain hotel got on a fuss with the waitress and encouraged by her manner he finally asked if he could come along to her room one night. She agreed and pointing out the door of her room – fourth on the left – said she'd expect him that night after everyone had gone to bed. Waiting until all was quiet he sneaked along and coming to her room – fourth on the left – he gently tapped on the door.

"Are you in, Honey?" he called in a semi-whisper. "Are you in?"

There was no response so he tapped again and called a little louder.

"Are you there, Honey? Answer me Honey."

He got answered all right, but it was by a gruff male voice.

"What the hell do you want?" the man snarled "This is a lavatory. Not a bloody beehive."

That name stuck with Clive Richards, yet he was as tough as they came – a good rugby player, cricketer and boxer. With a chap like him a sissy nickname just didn't matter. Unfortunately Honey lost his life in Egypt in the big brawl only a few years later. I heard how he died; he showed his courage and determination right to the end. He was a great bloke.

Although I always played a very cautious game, that old sinner Gandhi also added me to his list of victims – not that that would take much doing. Reg Ferguson and I had taken the annual draft of sale ewes down to the Sheffield ewe fair and on this occasion I was riding Gandhi. This is a five-day drive and for the whole of that time we just dawdled along behind the sheep, half asleep and rolling in the saddle any old how. Coming home we did the whole distance in the day – the best part of 50 miles and far enough to take the nonsense out of any horse. Yet when only about ten miles from home that old sod of a horse must have thought the time right. The place was, anyway, for I landed fair in the middle of a good clump of matagouri. The funny thing was that on the way down he'd had dozens of chances yet he'd never flicked an ear. You just never knew with that horse.

Jock McArthur – whom I mentioned earlier – came to Mt White to do some wild cattle hunting about that time and he was given Gandhi to ride. The old so-and-so was then at about his best – or worst – and how Jock fared with him I wouldn't know. Of course old Jock was a leading New Zealand rough rider and if Gandhi had dumped him he'd never have let on to us anyway. Jock was on that job for several weeks – on his own – and although I couldn't imagine a worse horse for that sort of work he got a lot of those cattle out. My bet is that Gandhi got him too.

Next to Gandhi as a chronic no-hoper I'd rate Phar Lap – or, as he became known, Fart Lap. About that time a young fellow came to Mt White as packman. As he'd not had much experience with horses he was given Phar Lap to ride. Apparently this boy had never heard of the famous Phar Lap. He must have thought his horse's name was Fart Lap. This was quite understandable as the old horse, always more dead than alive, had to be fairly kicked along and a habit he had was to let a real rouser go every time he was thumped in the ribs. So Fart Lap it became and Fart Lap it remained.

For all his docile ways old Fart Lap had one very disconcerting habit – if his rider got abusive with him, and that happened a lot, the old so-and-so would run backwards. He'd do this regardless of the going – on the hill, on a steep narrow pack-track or anywhere. This could be pretty frightening at times and I've seen some flying dismounts that would be worthy of a circus rider. But we got some good laughs out of that horse, and that's something.

Tan, the big raking chestnut, was just so ordinary that he's hardly worth a mention. Tom was even worse; his only claim to fame, as far as I could see, was that he was an even worse leader than the Binnie's Grey. It could be, also, that he might have qualified as the worst horse in New Zealand to sit on – he'd be on about a par with a log of wood.

Old Tom, poor old fellow, came to a sad end and what's more it nearly cost Honey Richards his job. The boss, Honey and myself had been over on Lower Pakety looking for hoggets that might have been hung up in lawyers. On the way home, late in the afternoon, we had to come up a steep, shady terrace. We'd had a string of hard frosts and with the track frozen like concrete it wasn't the best of going for horses. We were walking and leading our horses up this track, but only a chain or so from the top Honey's horse, old Tom, slipped on the icy ground. Honey hadn't a hope of holding him and all we could do was stand and watch him somersault right to the bottom a hundred feet below. Just as Tom was about to crash onto the flat Honey sang out, "Sit down there Tom."

Tom sat down alright – and it was for keeps. He had broken his back. Old Bully was hellish upset and I thought he was going to sack Honey on the spot. Fortunately we'd taken rifles with us – for pigs – and we were able to put the poor old fellow out of his misery in quick time. The boss made Honey do it too. It's never much of a job, shooting a horse.

Yet strangely enough although the hacks at that time were a pretty sorry lot Mr White had the best pack-team I've known. They had five of them in work and every one was a good 'un. There were two blacks, Johnny and Nugget, a lovely looking chestnut named Ginger, a big bay, George, and an old grey gelding named Beau, who was kept mainly as a spare.

Johnny, a nuggety half-draught, was the personality horse of the team. If ever the term could be applied to an animal he was a delightful character. Just as knowing as a man, he was as mischievous as a child and in the hands of an inexperienced horseman he could be a nightmare. The pack-track from the Mt White homestead out to Lochinver, where most of the mustering and sheep work is done, winds round the shores of beautiful Lake Letitia and if Johnny thought he could get away with it he'd suddenly wheel out into it and get his load soaking wet. Going through bush he'd deliberately leave the track and squeeze between two trees not quite wide enough for him and his load. The bread or whatever he was carrying would get crushed beyond repair. Another trick was to duck into a patch of scrub and hide. At times the packman, dawdling along behind his horses and half asleep, might be several hundred yards past before he realised Johnny was missing.

Pack-horses are turned loose and driven ahead of the packman, and this gave Johnny plenty of chances to get up to all sorts of capers.

But with anyone who knew horses and would stand no nonsense Johnny was a cracker. You could put anything on him – he'd have carried that Woodstock pig without batting an eye – and when loading him you could climb about on the top of him at will. When strapping on the "top load" the packman clambers up onto the horse's rump so that he is better able to adjust the load; some horses don't like this but with Johnny you simply grabbed hold of the crupper, put one foot on his hock and swung straight up. All these things help. With all this he was strong, not too high and had a beautiful balanced walk.

But perhaps his best feature as a pack-horse was that if there was even the slightest sign of an imbalance in his load he'd stop dead and wouldn't move until it was righted. An unbalanced load is the major cause of sore backs – one of the bugbears of a packman's job. Packmen don't usually carry scales with them – some do – but just rely on "hefting" the loads to get them balanced. It's easy to be a few pounds out, and that's enough to have a saddle listing to one side. Some horses don't like being caught once their heads are turned for home and a horse like Johnny that will stop himself is a real asset.

Each horse has its own place in the string and Johnny always took the lead. Nugget came next, then Ginger. George, as the youngest and a new-comer to the team, was forced to stop in the rear. This obviously irked him, but none of the others would permit him to take their place. Whether he liked it or not he had to trail along in the rear all day. Johnny as the leader made the pace, and when climbing it could be left to him entirely; every few hundred yards he'd stop for a spell and after a minute or so move off again of his own accord.

In good hands he was a cracker – you couldn't get better – but, boy, would he take advantage of a new chum. But it was pure mischievousness and I've always had a soft spot for that old horse.

The other three were all good pack-horses too, Ginger in particular being a beautiful, level-headed horse. Beau, although getting on in years, was one of those real old reliables. Old Con Hodgkinson, then one of the best-known deer hunters in the length of the island, put in a season at Mt White, shooting for the Department of Internal Affairs – who then controlled deer culling activities. When they finished up he carried on shooting for skins on his own account. Although as tough as they come, Con was getting on in years then and the boss let him take old Beau as his pack-horse. Beau was no chicken either, but Con – who was then getting past lumping skins – had him over every inch of that Lochinver country. He took him right out on the high tops and everywhere it was possible to take a horse. He reckoned old Beau was the best pack-horse he'd seen in his lifetime.

When shooting on good deer country a man's tally is determined by the number of skins he can carry – not by the number he can shoot – and having a horse to do the carrying makes all the difference. Old Con put in another 12 months at Mt White, shooting for himself. During that time he got a further 1200-odd skins. On top of that he'd just shot 800 for the department. Actually the four departmental men accounted for something over 3000 that season on Mt White alone – so that was well over 4000 all told off one place.

Even so these figures are far short of record tallies. In the Wanaka district some years later the two Erceg brothers accounted for over 2200 each in the seven-month shooting season – a record tally for two men. However they were only taking the tails – easier by far as it eliminates the back-breaking job of carrying skins. Con's tally of 1200-odd skins in 12 months wasn't bad, especially as he was working on country that had already been shot over for the previous six months. It's interesting to note also that Con purchased his tucker from the station and that his tucker bill for the 12 months was only £12 – $24. It seemed that he lived on venison and camp oven bread. All he had to buy was tea, salt and flour with, perhaps, a bit of butter now and again. How that old bloke stood up to that work, I wouldn't know. He owed a lot to old Beau, that's for sure.

But old Beau had one unusual disability – he was stone-blind in the dark. I was inclined to take this with a grain of salt until one night I had to ride him out to Nigger Hill, one of the mustering camps. I left the homestead in the late afternoon, in daylight, but I had a three-hour ride and during the evening a dirty southerly storm came up. I was then only halfway and in no time it was as black as the inside of a cow. At this stage the track wound in and out of three little gullies, in dense bush – known as the Scenic Railway – and as I had no hope of seeing the track I had to leave it entirely to old Beau. Well that horse was blind alright – stone-blind – and before we'd gone a chain into the bush it was obvious that he'd go over a bank and break his neck – and mine – if I attempted to ride him. It was an unusual and unnerving experience.

There was only one thing for it – get off and walk. Although I couldn't see an inch I was able to feel my way, but even so old Beau was so nervous that I had a job to get him to follow me. I've often ridden in pitch darkness and although at times it's not very pleasant most horses have no trouble following a track. Riding a blind horse was something I've never experienced before. Once was enough too.

There was quite a bit of cattle work at Mt White – they were then running about 500 head – and this gave me ample opportunity to show Jimmy off. I was there just before Jock McArthur's time and there were wild cattle on both the Lochinver country – at the head of the Esk – and up the Poulter. We occasionally had a go at them and although we were never able to yard them we had some mad scampers and a lot of fun. On one occasion Jimmy Searle and I were flat out trying to head some off on the Riversdale flat when a little terrace which I hadn't spotted suddenly loomed up half a chain ahead. It was only about a two-foot drop but we hit it at full gallop and the jolt, when we landed, would have sent most horses crashing head over heels. I thought we'd had it, but although it really staggered him Jimmy managed to keep his feet.

Jimmy Searle, who was a chain or two behind me, managed to pull his horse in. He was riding his own hack and it also was a pretty useful sort of a prad. A dark brown gelding – I can't remember what he called him – he was a slashing gig horse, especially on a journey, and was well known in that locality. Jim came from Springfield, a foothills township which like most such places in those days had always been a stamping ground for musterers. Over the years it has produced many who made their mark in that calling. Snowy and Ted King; Jim and his brother Les; Reg, Jim and Brig Ferguson; and Bill Nimmo. Then there was the Joines family – Harold, Bill and Arthur – and the latter's three sons, Stan, Harold and Frank. Between them they gave many years of service to the high country. But perhaps the best known of all was Harold White, who for 50 years or more has been a top New Zealand dog man.

They'd put in the winter down country and then in the spring roll their swags and travelling per pig-skin or by horse and gig head away up the gorges for another season on the high tops. Each man had his own horse and invariably it was a good one. Horses, and dogs, were a part of their life.

# 11. Black Jack

AFTER Con Hodgkinson's success at "skin hunting" I decided I'd have a go at the game myself. The following spring I rolled my swag and headed for the Rakaia Gorge. I'd got permission to shoot the Wilberforce, Harper and Avoca country on Mt Algidus and Glenthorne and Lake Coleridge stations. At that time "deer culling" had hardly got under way in Canterbury and this was then some of the best deer country in the island.

However before I started on that job I gave a hand with the shearing muster at High Peak, a grazing run in the mouth of the gorge. It was while I was there that my hack Jimmy put up a performance that for sheer guts would take a bit of beating. High Peak had a few young horses running out on the hill and Ted Relph, the manager, decided that it was time they were run in and handled a bit. One filly, somewhere about two and a half years of age, proved to be pretty wild. Every time we got the mob near the yards she'd make a break and take off. We finally yarded the others and Ted determined that if it took a week we'd get that filly – and what's more, we'd break her in while we had her.

Anyway after I'd given Jimmy a good spell I took off after her. With never a let-up I simply kept hard on her tail, following her wherever she went. I can't remember how long it took, but Jimmy never faltered; I kept at her without respite and we just wore that filly down. In the finish she was down to a trot. When I finally got her into the yard all her fight had gone and it was no trouble to get a rope on her. I might add that although she couldn't jump she several times floundered over fences whereas I'd have to ride some distance to the nearest gateway; a further big disadvantage was that my hack would be carrying the best part of 12 stone all told while she was as free as the wind. I reckon that was a pretty good performance of Jimmy's.

Yet this was not surprising, really, as it seems that if he persists a mounted man will always wear down a riderless horse – provided it's on fenced country and he has a decent hack. For the first few miles he'll be left far behind but in the end the riderless horse will usually toss it in. I've seen it more than once.

Anyway at Ted's suggestion I started handling that filly before she'd really had time to recover, and for all her previous wildness she gave little trouble. In no time she was responding well and looked the making of a nice sensible hack.

It so happened that when the shearing was due to start, a day or two later, only two of the six shearers turned up; this wasn't so good as there were a lot of sheep on hand – which couldn't be held for long – and the upshot was that I took on a stand until the rest of the gang arrived. This meant that except in the evening I wouldn't have much time to play around with the filly; to overcome this I used to tie her up alongside the shed. Every break I'd slip out and handle her. She eventually became so quiet that before I ever rode her I could climb all over her and she'd never flick an ear. Without ever using a rope I could handle both back and front legs at will; at times I tied sheets of newspaper on her mane and tail and even that didn't upset her. When I finally rode her she never even arched her back – except once – and she was soon going like a train.

But there was "that once". On that occasion I led her well away from the yard before I mounted and just as I was swinging into the saddle something upset her. Before I could get properly seated – that's my story anyway – she sent me flying and took off back to the shed. Not wanting the other chaps to know I'd been dumped I did my best to catch her before she got back. But there was no show. I was Queensland Harry to those blokes after that. But that was the only time she looked like doing anything and I thought she was right.

Actually I did have another spill off that filly but it was in no way her fault. I was playing around with her out in front of the shearers' hut – just showing off really – and with one of the blokes holding I sprang on her back. As I landed she flinched away and I came a cropper right under her feet. With any other young thing I'd probably have got my head kicked off. I could only hope I didn't look as silly as I felt.

When I look back I've been a bit lucky on several occasions over the years. On one occasion at Castle Hill – when I had Torchy – he came down with me when I was after cattle. When he scrambled to his feet I was unable to kick my feet clear. One foot was hung up in a stirrup and he took off with me dragging alongside. Fortunately I had safety clips on my saddle and after a few chains the stirrup leather pulled off.

I was on my own at the time and but for those clips I'd have been a goner. It wasn't a very pleasant experience.

I had a somewhat similar experience at Lake Coleridge soon after that. I was coming down a steep little pinch when the mare I was riding gave a sudden heave and I went flying. I was using an old military saddle with small stirrup irons and this time both feet stuck in the stirrups. I found myself hanging down over her withers in a pretty impossible position. Luckily for me one of the other chaps happened to be right alongside and he managed to grab the mare's head before she got going. I was always a bit nervy with horses and those incidents didn't help much.

However I did learn something from that High Peak filly. When I left she was taken over by one of the shepherds who was inclined to be a bit impatient with horses and somehow or other he didn't get on with her. In the end she really went dirty with that chap and turned out to be a real problem. I realised then that although I'd spent a lot of time handling her and had her as quiet as a sheep she had never once been bested by a man. When it did come to showdown she thought she was boss – and made the most of it. I never did much in the way of horse breaking, but I didn't make that mistake again. At some stage a young horse must be bested by man – even if it has to be roped and dumped – until it does accept his authority.

I never heard how that filly finished up. She was a good cut of a mare and should have made a good hack; if she didn't it was the fault of the horse breaker (or the would-be horse breaker).

During the summer months I shot on my own – something which should really not be permitted in that class of country – and as I needed a pack-horse Jack Murchison let me have old Dick, the big trotting horse. He had been pensioned and his place in the pack-team taken by the big black gelding that Jim Binnie and I had broken in. Jim Ferguson was mustering at the Lake that season and he had taken over the bay cob. In his hands he'd turned out to be a good horse; he was sturdily built and not too high; he was an ideal type for a shepherd's hack. On several occasions Jim had received pretty tempting offers for him. Black Jack was still on the place but his reputation was such that no-one was interested in him.

Apart from his peculiar aversion to carrying a rifle old Dick was just the horse for me – for all his age. This was in the summer of 1939, just 12 years after I first went to the Lake and as he was even then an aged horse I reckoned he'd now be round about 25. But he was still as game as ever and at times carried some pretty hefty loads, particularly when I was shifting camp. On those occasions he'd have all my gear – bedroll, tucker and cooking utensils – as well as any skins I had on hand. At times there'd be anything up to 20 dried skins, as well as ten or a dozen green skins. A fair sort of a load, all told, but the old bloke never faltered. He was a wonderful old horse.

Honey Richards joined me in the winter. As we then needed a second pack-horse the boss asked us if we'd take Black Jack and see how he shaped at the packing. I wasn't too keen, but feeling that Jim and I had fallen down on the job when we broke him in I could hardly refuse. I expected all sorts of trouble but strangely enough he proved to be an ideal horse for the job. We found that we could pile anything on him – even dried skins – and he never batted an eye. Dried skins can be a pretty frightening load for a horse until it becomes accustomed to them. They are simply folded in half lengthwise, and as a stag skin can be five feet or more in length they reach almost from a horse's head to its rump. A load for a horse is two bundles of 25 or so, one on each side. At times you can hardly see the horse for skins. They smell a bit too – especially stags – and there's usually a bit of fun when they are first loaded onto a horse.

Another point in Black Jack's favour was that he never attempted to stray as long as Dick was about. It was not necessary to tether him at night – if we'd left old Dick loose he'd have taken off for home in a flash. Yet for some reason Black Jack would never accept a rider. Once I tried him out. After we'd been packing him for some months he became so quiet that one morning, when I was bringing him in, I hopped on his back. I stopped there about five seconds.

However when Bill Russell went to the Lake just after that, breaking in horses, he took Black Jack in hand. Although he had to use hard measures, it seemed that he finally civilised him. That horse would have been eight or nine years old then. The last I heard of him was that he was sold at an Oxford horse sale along with a draft of Lake Coleridge young'ens. Black Jack was a good illustration of how easily a horse can go wrong.

That winter – 1939 – was one of the bad ones. Honey and I shot the Wilberforce country most of the time – with the Moa Hut our headquarters. For weeks on end everything was under snow. We had to pack out chaff for our horses but even so they had a pretty lean time. The Wilberforce – a sizeable river – and all creeks were frozen solid and we had to melt snow over the fire for all our requirements. This included water for the horses, and as 6 to 10 inches of snow is equal to only 1 inch of water it takes a fair bit of snow to give a horse a drink. However there was plenty of snow.

Under these conditions we never got many deer, so our horses didn't have much to do. But no skins means no money, and that winter cured me of deer shooting for a crust. As it happened there was a job offering at Grasmere Station at Cass – my old stamping ground – so in the spring I rolled my swag once again and was back in the mustering game. I started at Grasmere on 3 September – the day that Britain declared war on Germany.

Jimmy had been turned out at the homestead during the winter, so he was pretty soft. That ride through to Grasmere proved to be the severest test he'd yet had. From the Lake homestead through to Lake Lyndon – a distance of about nine miles – the road follows up the Acheron valley and for most of the way it climbs steadily. The river, which is a branch of the Rakaia, drains due south and in a bad winter the valley is a real trap for snow. I was well aware of this, but as there wasn't a skerrick of snow on the flats at the Lake homestead I never for a moment thought that I'd strike more than a few inches on the way through to Lake Lyndon. I was in for a rude awakening. By the time I was halfway the snow was up to my horse's belly – and getting deeper all the time. At one stage, when the road sidled round a dark face, Jimmy floundered over the bank and with the snow then up to the saddle flaps I thought we'd had it.

I didn't dare get off and lead him, as in that depth of snow I'd have been unable to keep clear of his lunging and plunging and could easily have been tramped under foot. One alternative would have been to take the saddle off, leave Jimmy to it and battle my way back to the Lake on foot. However although I'd have had our tracks to follow I didn't fancy that idea and in the end I stuck to the ship. We finally got back up onto the roadway but it was a tremendous struggle and only a horse with guts and courage to burn would have made it.

From then on the going wasn't so bad, although another problem was a gateway just before we reached Lake Lyndon. The snow was then about two-thirds up the gate – 30 in or so – and as there was no show of opening it I had to lift it off its hinges and drag it out on top of the snow. A 12-foot, five-barred gate is pretty heavy – it seemed a ton to me – and that was another big struggle. I didn't attempt to put it back anyway.

However we eventually got through to the main highway to find that the grader had been through and that our troubles were over. How long those few miles took us I can't remember. We were only in snow, of any depth, for about five miles, but at its worst it was about up to the saddle flaps. Thats a hell of a lot of snow for a horse to plough through. To make matters worse it had been lying most of the winter. Wet and soggy, it had packed hard. From the Lake homestead to Grasmere is only about 30 miles – normally an easy day's ride – but I was very pleased to dig in at Castle Hill that night. I'll bet Jimmy was too.

I discovered later that the level of Lake Lyndon – 2769 ft – is almost exactly 1100 ft higher than Lake Coleridge. But even so an increase of just on 3 ft in the depth of snow was quite surprising. That '39 winter cured me of snow anyway.

Grasmere Station was then owned by McLeod and Orbell. David McLeod, an Englishman who came out to this country in the mid-'20s, put in a season or two mustering in the Canterbury high country and was so taken with the life that in partnership with Leslie Orbell he finally purchased Grasmere.

I first met him at Lake Coleridge about 1928, when he passed through Lake Coleridge on his way to Craigieburn from Raincliff in South Canterbury. His hack, which I remember well, was a big rakey mare of a most peculiar colour – a grey which, of all colours, was tinged with an unusual shade of pink! However although only young she seemed a good sort of a hack – even if she had dumped him in the middle of the road that very day. David was heading through to Lake Lyndon and I gathered that he, too, had good cause to remember that particular stretch of road.

About 12 months before he'd come through from Craigieburn in a gig, and at one deep little gully he'd found the road completely blocked by a slip a chain or so wide. With no shovel or tools of any description he couldn't do much about it, but he finally decided he'd unyoke his horse, take all the gear out of the gig, and see if he could pull it across himself. However the grade was too steep and the gig finally topped over and rolled right to the bottom of the gully. He had to ride his mare through to Lake Coleridge for assistance.

When I went to Grasmere 10 or 12 years later the "pink" mare Dinah, was an iron-grey, the pink "undertone" having disappeared with age. From a leggy somewhat ungainly youngster she had developed into a big, strong, impressive looking mare – a good type of station hack, except that you needed a stepladder to get on her. She was also a pretty good jumper and David seldom had to open a gate when riding her. She could also be led up to a fence and from a "standing start" hop over it with ease. In places where it wasn't possible to ride her at a fence this was very handy.

Dinah was bred from on two occasions while at Grasmere and both foals turned out to be attractive greys. The first one, a colt, was sold while quite young and the other, a very attractive filly, had the misfortune to break a leg. Reg Ferguson – who had the Bealey pub for some years – had her at the time. During the day he tied her to a scrub bush while he went looking for some sheep that had got stuck in a gully. While away something must have startled Wanda – as she was called – for she pulled back, broke the reins and took off down hill at a hand gallop; unable to keep her feet she finally crashed and with a leg badly broken had to be destroyed.

The other station hacks during my time at Grasmere were a big bay mare called Sybil, which although past her best was still a grand hack, a white mare, Betsy, and a classy little chestnut mare named Rata. Like Dinah, Betsy was a bit of a jumper, her worst feature being that no paddock would hold her; if she felt inclined she'd wander anywhere. In her day Rata had been one of Leslie Orbell's polo ponies, but when the game was in the doldrums during the Depression, she and several of the others had been sent up to Grasmere. When they were sent back she had been retained as a station hack.

Sybil, who was a useful harness horse, was often used in the gig to get the mail and any odds and ends that were sent up to Cass by rail. I remember this in particular for an incident which – as it turned out – caused a fair bit of amusement.

At the time the station cowman was away on holiday. I had undertaken to do his job – milk the cows, do the separating, get the mail – and this of course gave me the opportunity to get down on a bit of cream. I used to visit one of the couples over at Cass now and again and as I knew that I'd be going over there for the mail that day I decided that I'd fill up a 2 lb jar with cream and sneak it across. It so happened that I had a pair of gumboots to return to them and in order to escape detection I slipped the jar into one of those boots. Anyway I duly yoked old Sybil into the gig, collected the gumboots and set out for Cass. However as I was passing the sheepyards the boss came across and said that he had occasion to go there himself and that I need not bother. Spotting the gumboots he asked what they were for. I told

*Above:* "Some of their horses became as knowing as a man." Ready to start on the muster on the Waipaoa Station near Gisborne. *N.Z. Herald*

*Below:* Pony club riders cooling off in the Ashley River during a trek into Lees Valley. *Mrs B. Page*

*Above:* "The St James pack-team was a common sight in the Hanmer township." This time the horses are loaded with deer skins. *H. Riddle*

*Below:* Amateur blacksmith Harvey Riddle, one-time manager of St James. *H. Riddle*

*Above:* The St James pack-team coming out over Fowlers Pass. *H. Riddle*

*Below:* "Predominantly a dark brown, they were low set, nuggety horses, ideal for pack work." The type of horse bred at St James in the 1920s. *P. Newton*

*Above:* St James horses today. A horse sale in the 1970s. *J. Stevenson*

*This picture:* "At one time – before the advent of road transport – it was commonplace to see big mobs." Six thousand Winterslow sheep on the road in recent times. *D. Wightman*

him, saying that there was no hurry and that I'd deliver them next time I went across; no need for him to bother with them. But he insisted that as he was passing the house in question he'd deliver them himself. As I had no appropriate answer to this away he went – boots and all.

I quite expected a few awkward questions when he got back. But although I thought I wasn't bad at the bull I just couldn't think up a story that would meet the occasion. But luck was with me. Apparently he never suspected a thing and he actually delivered a jar of his own cream without knowing it. You had to be pretty smart to put one over David McLeod too.

As far as I know he never found out about that cream, but it's quite on the cards that he'll read this story of mine sometime. However that was nearly 40 years ago and I don't think we'd have words over it now.

At Grasmere the ewes were run on easy hill country in the main, and it was the practice to ride round them during the lambing. With little to do apart from picking up the odd ewe that had got cast while trying to lamb this could get fairly tedious – just ride, ride, ride over the same stretch of country day after day after day. The lambing lasted about six weeks and on a lifeless horse it got pretty monotonous, to say the least. However it was at this job that Jimmy was at his best. Full of life and spring, he was lovely to sit on no matter how long the day and he was just as free at the end of the lambing as he was at the start.

It was a pleasure to ride him and even that lambing job had its compensations. With a good hack it can be very pleasant riding round that sort of country in the spring of the year.

Another memory of those days at Grasmere is that it was there that I first had the pleasure of meeting "old" Freeman Holmes, that grand old man of the trotting game. He and Clarrie Rhodes, another well-known trotting enthusiast, used to come up that way deer shooting. As they sometimes made the Bealey Hotel their headquarters they came well known to the local boys.

Freeman, who even then was well into his 60s, was still a leading reinsman and he went on to establish a record as a trainer, driver and rider that must surely be unique in the racing game. Starting off as a stable boy when he first left school he rode his first major winner when he piloted Manton to victory in the New Zealand Galloping Cup in 1888. He was then just 17 years of age. He had his last win with Suspense at Greymouth in 1944 at the remarkable age of 73. During that period – close on 60 years – he proved to be a trainer and driver second to none.

Among the more notable of his early successes was that of riding Liberator to a win in the Grand National Hurdles in 1897. His first win behind a trotter was with Black Oak at Ashburton in 1901. As owner, driver and trainer he also had the distinction of winning the 1919 New Zealand Trotting Cup with Trix Pointer – a mare that he himself had imported from the States. But what was probably his greatest achievement as a driver was that of winning the first Inter-Dominion Trotting Championship with Evicus at Perth in 1936. Two other notable successes were those of driving the winner of the Auckland Trotting Cup in two successive years – Roi L'Or in 1934 and Graham Direct in 1935. As if to prove his versatility he also trained Minerval, who won the Auckland Galloping Cup in 1933.

But perhaps even more remarkable are the successes of three of his sons, Freeman Junior, Maurice and Alan. Emulating their illustrious father, all three of them have had the distinction of driving the winner of a New Zealand Trotting Cup – the ultimate in the trotting game in this country. Maurice, the most successful of the three brothers, has the honour of holding the New Zealand record for winning drives. This stands at the incredible tally of over 1500 – just on 500 more than any other driver.

In this respect he leaves his father well behind, but for all his prowess it is surely "old" Freeman who must be counted the doyen of the New Zealand trotting world. Freeman Holmes died in 1967 at the remarkable age of 96, but as long as we have horses his name will live on. He was a wonderful man and a credit to the game.

Yet another member of that famous family – Freeman the third – has established himself as a trotting personality of note. As well as being a prominent trotting commentator he has his own training establishment at Springston and is fast proving himself to be in the top flight. Of his many achievements in this field the most notable will be his remarkable run of successes with the Batchelor Hanover colt Noodlum. Noodlum suffered a breakdown in 1976 after establishing himself as probably the greatest two-year and three-year-old this country has produced. Whether he ever races again remains to be seen, but his success on the track was such that he is now in great demand for stud purposes.

It was Noodlum's successes as much as anything that really put "young" Freeman on the map. It seems that he too will add lustre to what is one of the great names in the history of New Zealand trotting.

# 12. Mainly of mules

WITH many of my mustering mates I volunteered for overseas service at the end of the 1939-40 season, but as high country mustering was considered to be an essential occupation it was another two seasons before I was called up. When I did get the word no time was wasted and I was on the water bound for the Middle East in only a few weeks.

I left Jimmy and a couple of dogs at Grasmere and I thought this would be the end of horses and dogs for a year or two. As far as horses were concerned this was far from the case. What with Arab studs, gharry horses, mules and German horse-drawn transport (in the latter months of the Italian campaign) horses were still to play an important part in our activities.

We left New Zealand in the *Mooltan* and apart from a brief stop at Perth I spent the next few weeks hanging over the ship's side gazing at the water. It was not until we pulled into Aden, at the entrance to the Red Sea, that anything happened to break the monotony. This was my first introduction to the "Wogs"; in little craft of all descriptions they swarmed round our boat offering all sorts of fruit and drinks and diving for pennies which our boys tossed into the water. They were pretty good at this and I don't think a single coin was lost.

For some weeks we were based at Maadi camp, a few miles from Cairo, and needless to add we soon made our presence felt in that most intriguing of cities. Of particular interest to me were the strings of gharries which lined the curbs, every here and there, as they waited for customers. Egypt, of course, is virtually the home of the Arab horses – the most artistocratic of all breeds – and I had imagined that even among the gharry horses I'd see some fine animals. But this was not the case; in the main they were a pretty ordinary, sorry looking lot. This was not surprising really for as we saw more of that country we wondered how on earth they were able to feed them. But they served the purpose, and we Kiwis made full use of them – often to the sorrow of their owners.

While on the subject of gharries there was one incident – it happened in the middle of Cairo – that caused a lot of fun. While a driver was assisting a couple of very elegant ladies into the gharry I slipped up on the blind side of his horse and uncoupling one rein from the bit ring attached it to the chin strap. After seeing the ladies safely seated the driver whipped up his horse and took off at a spanking trot – quite unaware that anything was wrong. With only one rein attached to the bit he was unable to control his horse at all; all he could do was pull it one way. He finally finished up on the footpath with horse and gharry jammed up against the front entrance to a shop. The abuse of the shopkeeper and the pedestrians, the screaming of the two passengers and the curses of the driver made it as good as a circus. However no-one was killed and I'll bet those two ladies look back on that incident as the thrill of their lives.

There were two occasions, however, when we did see some beautiful horses in Cairo. One was a visit to the police barracks where they kept and – if I remember rightly bred – the police horses. The other occasion was a visit to a leading Arab stud.

The police horses must have been bred from high class Arabs, for they were mainly whites or greys, and there were some magnificent animals among them. They were housed in an impressive castlelike establishment within the city, and were obviously the pride and joy of their handlers. Our visit to the Arab stud was an organised trip and was – or should have been – one of the highlights of my sojourn in that country. Unfortunately my recollections of that outing are rather hazy, as on the way out there we got hold of a bottle of a particularly potent drink known as Zib-bib. The taste wasn't that good – something like kerosene that had been filtered through an old sock – but we liked what it did to us. Having been mainly confined to a horrible brew of local beer known as Stella we made the most of it while the going was good.

As a result I can't remember much about that Arab stud. But it was an amazing establishment, owned, I imagine, by some Egyptian potentate. The horses – the stallions in particular – were truly magnificent.

When the North African campaign was finally wrapped up we were shipped over to Italy. Although they had got most of our money I don't think the Egyptians were sorry to see us go. The Kiwis – known as Freyberg's 40,000 Thieves – had slightly different ways to theirs! But it was a case of rob or be robbed, and the Wogs took a lot of beating at that game. However just to show that we could hold our own in that department another bloke – I'll call him George – and I staged a last little stunt before we left. This story doesn't concern horses in any way but its a break from that subject and that might be welcome.

It so happened that I did a bit of hair cutting in our company and one day, just before this, my clippers disappeared. I don't know who pinched them, but I blamed the Wogs. Such being the case – if it was them – I reckoned that what was good for the goose was good for the gander.

Anyway George and I took off for a barber's shop. The scheme was that I would have a hair cut while he waited for me in the shop. Just as my hair cut was finished and I was stepping out of the chair George was to stroll across to the far end of the shelf where the barber's equipment was laid out and quite openly pick up a razor and put it in his pocket. As was always the case when there were Kiwis about, the shop staff consisted of the barbers themselves plus three or four others whose sole purpose was to see that nothing was stolen.

As we had anticipated, the whole crowd – barbers and all – immediately flocked round George demanding that he give the razor back. While this was going on I pocketed a pair of clippers. George gave the razor back with only a show of argument and every one was happy. Including me – they were good clippers.

When we finally landed in Italy it was getting on towards the winter. In no time many of the roads and tracks were such that it was virtually impossible to get supplies up to the front line troops by normal transport. As an alternative the big noises decided to use pack mules, and in due course they called for volunteers to take on the packing. Most of the high country musterers in our outfit – the 23rd Battalion – volunteered, but in the end only two were accepted: myself and my old Springfield mate "Brigadier" Ferguson. We were also given a mob of Italian muleteers. However that brainwave didn't work. If the Jerries did us over, and I sometimes thought they did so deliberately, the Ities would take off and leave Brig and I to it.

The heads then decided that they'd replace them with some of our blokes. These chaps came from all walks of life – bank clerks, counter jumpers and insurance wallahs – and some of them knew as much about pack mules as I do about a knitting machine. However they did stick to the ship when the going was tough.

Each man had two mules, which he had to look after and feed. This wasn't so easy as the mules were simply tied to olive trees when not working; with not a vestige of grass about, the only tucker available was what dry straw the chaps were able to pinch from the surrounding farms. However we occasionally got a ration of oats for them and they survived. They say a mule can live on the smell of an oily rag – these mules of ours just about proved that. Our battalion mule team numbered about 30, and 30 mules get away with a fair bit of tucker. I'll bet the Itie farmers were always pleased when we shifted camp.

In the main our job was to take the morning and evening meals up to the men in the front line. Of course this had to be done under cover of darkness. At times the track was a foot deep in mud, and in the dark it could be slow, difficult going. Another drawback was that we had no suitable containers for packing such items as stew, porridge and tea; so if we got up there with half of what we started out with we were lucky – or the blokes were. At the start we were dependent on open containers without lids, and if a mule stumbled there was porridge and stew everywhere. On one occasion I gave the mule in front of me a slap on the rump to hurry it up when it gave a sudden lurch forward; its front feet stuck in the mud and the poor beggar went straight down on its nose in the slush. It was carrying stew, and there wasn't much of that lot left when it finally got to its feet again.

However we finally got better gear for the job, and by the time we finished with the mules we were pretty well organised.

The mules, so I was told, came from both Syria and Greece and were of two distinct types. Those from Syria were small and light in the bone, but the others were beautiful, big, strong animals. Bigger by far than any mules I'd seen. What did surprise me was that they were all beautifully quiet to handle. It's commonly said that a mule can kick the eye out of a needle, and for a start I was pretty wary of them. However there is always the exception – as one of our chaps found to his sorrow.

As I mentioned earlier, all the actual packing was done during the hours of darkness. After returning from delivering the evening meal this chap tied his mules to the usual tree, gave them a feed of straw and headed "home" to his bivvie. In the morning before daybreak he went across to saddle up, gave one of them a friendly pat, and the next thing he knew he got both heels fair in the bread-basket. Hearing his howls of anguish I went across to see what was wrong. There, tied to the tree, was a strange mule. It didn't take long to guess what had happened –

the Maori Battalion's mule team was just across the gully from ours and during the night they had sneaked across and changed one of their dirty mules for one of our quiet ones.

We had another bloke with us who wasn't bad at the poetry. When one of his mules stopped a bit of shrapnel one night he christened him Shrapnel and made up a rhyme to commemorate the occasion. I've had to break it down a bit, but it's still good old Kiwi language. Anyway, with apologies to the author – and readers – I've taken the liberty of recording it. Here's how it goes:

*The Mulies' Lament*

*I'll tell you a story of life in the raw*
*That happened to me in the Second World War,*
*Where 88s plastered us all through the night*
*As we lay in our slitties and shivered with fright.*

*This story I'll tell you is strange and true*
*But if you don't believe it you know what to do;*
*Like fools we had strayed from that haven of rest*
*A place in the Army that's known as B. Ech.*

*We had a two-pipper, a towney, in charge*
*And a bastard called Newton was acting as serg.*
*I will say of Newton he helped with the loads*
*But all the Big Chief did was get in the road.*

*Twice a day up the line with tucker we'd go;*
*T'was done in pitch darkness and the going was slow.*
*But those mules were hardy and bred for the rough*
*And give them their dues the bastards were tough.*

*And sometimes we sang as we plodded along*
*And in the next verse I'll give you our song;*
*It may not be high-brow and it also lacks class,*
*But remember we sang it in mud to the arse.*

*Get along Shrapnel, there's work to be done;*
*If we don't hurry up we'll be shelled by the Hun.*
*Both about buggered and ready for tea,*
*Wet straw for you boy, cold bully for me.*

*Get along Shrapnel – more effort, bud,*
*Though I know it's hard going in three feet of mud.*
*But we'll make it old fellow for the top is in sight;*
*I'm glad we just go to Headquarters tonight.*

*Thank goodness, old fellow, we're over the top,*
*For like you, old crow bait, I'm ready to drop;*
*I'll give you a feed as soon as we're home,*
*Then back to my bivvie – the place I call home.*

*Well here is the company, lads, unload old Shrap*
*And bolt down that tucker so we can get back,*
*The track is a bog from here back to camp*
*And even when dry it's a bloody stiff tramp.*

*How is the grub boys, it smells good to me;*
*There's plenty of stew but bugger all tea.*
*Shrap fell in a shell hole on the way to the top*
*And up went the dixie and out went the lot.*

*But don't call him a bastard, he wouldn't like that,*
*Far better to give him a nice friendly pat.*
*But don't touch that whopping big gash in his arse;*
*'Twas made by some shrapnel that went whizzing past.*

*Well cheerio boys, I've got to get back;*
*Old Shrap is fair starving and I'm pretty slack.*
*If we get back to camp our job's not been in vain*
*And at day-break tomorrow we'll be with you again.*

*But when Hitler is finished and fighting is 'oer*
*And I'm back in New Zealand to wander no more,*
*If pack mules are mentioned I'll say, 'Strike me pink,*
*'Twas those bloody mules that drove me to drink!'*

But we did get a bit of fun out of those mules, even if it was mostly at someone else's expense. On one occasion when the battalion was out resting we staged a race meeting, and although the performances of some of the mules left a lot to be desired, to say the least, it was a great day's entertainment. One of the chaps ran a "book", but as no one had the least idea as to how any of the "racehorses" would perform it was a real gamble for both the bookie and the betters.

The riding was all bare-back, and as none of the mules appeared to have ever been mounted it was almost impossible for the riders to control them. However a few of them performed quite well – often with a bit of help from those spectators who had backed them – but I doubt if some of them have finished the course yet. One mule which had been going pretty well suddenly decided he'd gone far enough when a chain or so short of the finishing line. In the end a crowd of enthusiasts who had backed it practically carried it over the line. Some of them dumped their riders smartly and one of them took off, flat out, in the wrong direction; it disappeared among the olive trees, its rider hanging on like grim death. About two races later it appeared again at the bottom of the straight and showing a fair turn of speed almost won that particular race.

All told it was a hilarious day for all concerned, although some of the riders had nothing but bruises to show for it. But it was better than packing, that's for sure.

It was soon after this that we had another experience with horses which was of particular interest to me. The Free French, who as our allies played a useful part in both the North African and Italian campaigns, had among their contingent a troop of mounted men, Moroccans I believe they were. I had first seen them in North Africa, but it was not until they joined us in Italy that I got the opportunity to have a look at their horses. Although a mixed lot – all colours – there were some splendid horses among them, and one in particular really caught my eye. A dark brown gelding with a heavy mane and tail it stood only about 14½ hands, but it was a beautifully coupled, sturdy little horse and with the appearance and carriage of a stallion. It was a striking looking animal.

I went across to have a closer look at it and at its rider – a black bearded, villainous looking character dressed in the heavy robes typical of the North Africans. Whether he could speak English – or any other language – I wouldn't know, for in response to my "Saida George" he simply fixed me with an unblinking, uncompromising stare. If looks could have killed I'd have dropped stone dead. I don't know just what part those troops played in the actual fighting, but I'd hate to have that bloke after my blood on a dark night. I wouldn't have minded his horse though.

We had another experience in Italy which was rather saddening to a lover of horses. At that time the Allies were right on top, although the Jerries – marvellous fighting men – still fought on with fanatical courage. They were then reduced almost entirely to horse transport and as part of war strategy it was necessary to destroy as many of those horses as possible. Our Air Force chaps were straffing them at every opportunity and towards the end of the campaign there were dead horses – and oxen – lying about everywhere. There were some beautiful horses among them and it was a sickening sight.

Many of them were simply abandoned alive, and at one stage the Maori Battalion was almost a mounted unit. They weren't going to walk when they could ride. They were catching these wandering horses and after rigging up makeshift harness they were doing it the easy way. Their "hacks" varied from big, lumbering draughts to the lightest of horses and were a queer looking lot. It's a sorry situation that hasn't got a laugh of some sort in it.

When the big brawl in Italy was finished I had several trips up into Austria (A W O L ) and on one occasion I saw several big mobs of horses running loose on the mountainside. We were told that they had been abandoned by the Cossacks – the famous Russian cavalry regiment – when they had been returned to Russia. I don't know the story exactly, but although Russia was on our side these Cossacks were actually fighting for the Jerries. The Cossacks are of course famed throughout the world for their horses and horsemanship, and they must have done it hard leaving their horses behind. I never got close enough to have a good look at them but there must have been some good horses among them. They were running in mobs of up to a hundred or more, semi-wild, and I don't know what eventually happened to them.

My main object in making those trips up into Austria was to get to Vienna in the hope that I could see those marvellous white horses of the Spanish Riding School. However I was never able to make it as the Russians were then occupying that part of Austria and there wasn't a hope of getting past their boundaries. Perhaps it's just as well I didn't as I might have finished up digging salt in the Siberian saltmines. But I'd have loved to have seen those horses.

I might add that when the New Zealand Division was stationed at Trieste at the conclusion of the campaign, some of our trotting enthusiasts got together and organised a properly conducted race meeting. Although it was run almost entirely by our chaps, the co-operation of the Italians was such that we were able to get enough horses to stage a full programme. Some of those horses were top performers, but it is interesting to note that practically all of them were straight-out trotters (although their blood lines were virtually the same as our own pacing horses). For some reason the pacing gait is not favoured in Italy and those horses play a very minor part in the sport.

I missed that meeting, which was held at the well-known Montebella trotting grounds, but by all accounts it was a great success. It seems that wherever we went horses just had to come into the picture.

# 13. Billy the second

IT seems that my "Cook's tour" of the Middle East must have knocked some sense into me, for soon after returning to New Zealand I got married – about the only sensible thing I've ever done. However before the big event another chap and I put in a few months deer shooting for skins at Mt White. Jimmy, my hack, was still at Grasmere so I took him in there and although he was then over 20 he was as game and as lively as ever. We used the station pack-team to cart our skins into the homestead and Jimmy was used mainly for carrying skins back to camp after a day's shoot. Even so he sometimes had some pretty stiff climbs loaded with green skins, and at times I wondered if his old heart would make it. He'd have dropped dead before he'd peak.

On one occasion when I went into the homestead to get the pack-horses the team was away on another job and the only horse available was a half-broken youngster, a big, powerful horse called George. I assured the boss that packing skins would be good education for him, and against his better judgment he let me take him.

We were camped at Nigger Hill at the time – about 11 miles out – and the next morning we roped George up and loaded him with two tremendous bundles of skins. When we had them securely fastened we unroped his legs and taking him firmly by the head I very cautiously led him away. As I remarked earlier, bundles of skins are a pretty fearsome load to put on a young horse. George's reactions were about what we should have expected. He gave an almighty bound. The bundle of skins on my side caught me fair behind the shoulders, knocking me flying, and George took off like a mad thing. With a crashing of branches he disappeared into a clump of bush along the track and when he appeared again, heading flat out for home, you could hardly see him or the skins for trailing branches.

We caught up with him at the first gateway and although he'd had a fair go at dumping them the bundles of skins were still intact. But young George was a pretty civil horse when we got into the homestead that night.

When we finished up on the shooting I took on a job that was a complete change from anything I had previously undertaken. This was a position as a farm valuer with the Valuation Department, and as I was to be stationed in Christchurch this meant that I would have no further use for Jimmy. The problem, then, was what to do with him. For all his age he was as keen as ever, but an inconsiderate rider might ride him into the ground. I wasn't going to risk this and although he was far from done I decided I'd have him put down. So I left him at Grasmere having got David McLeod's promise that this would be done. It seemed a harsh way to treat an old friend – for such indeed he was – but I knew in my heart it would be doing him a kindness. However it was a sad finale for a grand little horse.

For the next few years – as a collar and tie man – I had little to do with horses except for the odd times I borrowed a hack to ride round a hill country property. Indeed World War II virtually marked the end of an era with the horses; their place was taken by the tractor and the land rover, and it seemed that the day of the horse as the old-timers knew it was gone. For my part – horror of horrors – I even rode a push-bike to work on the odd occasions when I was spending a day or two in the office.

Although I was never really a racing man, Oscar Watson's Lady Ngaio was then at her best and was getting her full share of wins. As we were living quite close to the Addington trotting grounds I sometimes biked round to the railway yards at the back of the course and climbing up on a sheep truck – just as Terry Jones and I had done in our schooldays – watched her go. A bookie was still on the job and I picked up quite a few bob on that mare. That "Scotsman's grandstand" was pretty handy – you didn't even need to have a shave to go there, let alone wear a collar and tie.

There were two other horses of about that time which were to establish themselves among the all-time greats of New Zealand racing. They were Highland Fling and Johnny Globe. Highland Fling had the distinction of winning the New Zealand Trotting Cup in both 1947 and 1948, and in the same week of his second win he established a New Zealand time trial mile record of 1.57⅘. To cap this he came out again that same day and won the New Zealand Free-for-All. This surely places him in a class of his own – indeed some students of the game considered him to be the greatest pacing horse this country has produced.

Johnny Globe, then the idol of the Addington crowds, had his day a year or two later. He won the Trotting Cup in 1954 in world record time and he also had the distinction of breaking the world pacing records for the mile, the mile and a quarter and the mile and threequarters. He was a mighty little horse.

It was about this time also that I happened to be at Springfield – reminiscing in the pub – when Bill Russell called in. He was heading back to Lake Coleridge and the hack he was riding, an attractive grey gelding, really took my eye. A Coleridge-bred horse, it was actually out of the big blue mare that I mentioned earlier. For some unknown reason I finished up buying him. I had nothing for him to do and nowhere to keep him, but the beer was talking and I decided I just had to have that horse. Fortunately Bill offered to keep him until the end of the mustering season in three month's time, and by then I'd found somewhere to run him and someone to work him. What's more to the point is that I had a horse once again.

As it happened that horse – Bill I called him – turned out to be a lucky buy. The next summer I got six weeks leave of absence from the department and with the object of writing about the Canterbury high country stations I rode from Glen Lynn station at the head of Lake Ohau right through to Molesworth. I wrote of this experience in *High Country Journey.* All I need to say now is that Bill proved himself to be an outstanding hack. We stuck to the heart of the back country for the whole trip – somewhere about 500 miles – and with rivers to ford and ranges to cross it was a real test for a horse. Dependable, tough and above all easy to sit on, Bill was as good as they come.

Des Gregan, now with the Lands and Survey Department, accompanied me for most of the journey and he managed to pick up a horse from a Mackenzie Country rabbiter. A solid plodder, it had the constitution to stand such a journey and I've no doubt that in the end he'd have worn Bill down. But even so I can hardly count him as one of the great horses of all time. We christened him Concrete and with all due respects to Des, who became quite attached to him, that name suited him down to the ground.

I might add that on the way through we spent a couple of days at the Ada, the St James outstation. At the time all hands were busy mustering the horses and sorting out the sale draft. Fred Hampton, a well-known North Canterbury horse breaker, was in charge and his two offsiders were Ross Clark and Dave Haldane. I think this was the last time the St James horses were sent to Rangiora, as that sale was discontinued. For the next few years they were taken out to Culverden and sold in the railway yards. When the Jacks Pass road was put in, it was decided that the sale would be held on the St James homestead yards. This sale, which is held every other year, has become a very popular fixture.

It was while I was in Christchurch with the Valuation Department that Lance Skuthorpe, a famous Australian rough rider, made a tour of this country. With several noted outlaw horses and a team of top riders – including his sister Violet – he gave exhibitions throughout the country. His tour must have been a great success. His star horse was a big bay gelding called Aristocrat that was as savage a bucking horse as I've seen. Skuthorpe offered good money to any local rider who could sit him. Another particular vicious horse was a mare named Baxters Bay; she also could test the best of riders.

I went to his show twice, as a spectator I might add, once at Christchurch and once at Blenheim. Several riders tried their skill on these two horses but no-one looked like sitting them. At Christchurch I witnessed the shortest ride I've seen. The horse – it was Aristocrat – was held in an enclosed shute until the rider was mounted and when he was all set a gate was opened into the ring. When the gate was opened on this occasion Aristocrat gave one almighty heave and his rider shot into view and landed almost in the centre of the ring. A second or two the horse followed, head down and really going to town.

I was told that Queensland Harry, then an old man, had a go at Aristocrat in Christchurch; he did not ride him, but he claimed that he was pulled off by the flank-rope so he followed the show down to Ashburton and had another go. He didn't quite see the time out, but he put up such a great exhibition that Skuthorpe paid up.

Although few local riders were able to best his two star horses, Skuthorpe and his riders gave some grand exhibitions of riding. Dick Skinner, his right hand man for many years, was particularly good and Violet Skuthorpe, who also gave exhibition rides, could sit a horse with the best of them.

The Skuthorpe name is a household word throughout Australia, especially in the outback where a good rider takes second place to none; however it was not the Lance Skuthorpe that we New Zealanders know that really captured the imagination of the Australian people. This honour must go to the original Lance (L. A.) Skuthorpe, the father of "young" Lance and Violet. For all his prowess on a horse young Lance never quite achieved the fame of his famous father, although he and his sister travelled the world. "Old" Lance, a typical Australian bushman, must have been a remarkable personality; a born showman, a bare knuckle fighter who would never admit defeat and probably the most famous rider his country has known, he was one of the few who became a legend in his own lifetime.

As a rider he was at his peak about the turn of the century when still in his 20s, but it is claimed that he was 80 years of age before a horse ever bested him. He died in 1958, aged 87.

What was probably his most memorable riding feat was his re-enactment of Adam Lindsay Gordon's spectacular and extremely dangerous jump at Mt Gambier, in South Australia. This involved jumping a 4 ft 6 in fence erected round the edge of a cliff and only two horse lengths from its edge. The horse had to pivot as it landed – virtually turn while in the air – or both it and its rider would crash to their death on the rocks below. It was a crazy undertaking and only a rider with unlimited confidence in himself and his horse – a man in a thousand – would ever think of it. Lance Skuthorpe made it, although with only inches to spare. It is said that such was his tension that he had to be lifted from the saddle. The story of that jump is graphically recounted in Jack Pollard's book *The Horse Tamer,* the story of Lance Skuthorpe's life.

Adam Lindsay Gordon, who later achieved fame as a poet, was once a mounted policeman. He was also a superb rider. A monument to commemorate his fantastic jump – known as Gordon's Leap – was erected nearby. It is said that Gordon did the jump in response to a taunt. It earned him undying fame. Lance Skuthorpe did it for a £10 bet and his horse, an ex-steeplechaser named Wallace, was borrowed for the occasion.

After my ride through the back country I got the urge to take to the hills again, and soon after that I took on a job on Nokomai Station in Southland. Nokomai, then some 80,000 acres and carrying 20,000 sheep, took in the north-western side of the Garvie Range from the Mataura river – at the homestead end of the property – to within a mile or so of the old Nevis Hotel. A full 20 miles. Although the country runs up to over 6000 ft the mustering of all the back-country is done on horseback. The ewe blocks, strangely enough, have to be mustered

*Above:* Mrs Diana Bain on Hunters Moon. *M. F. Hooper*

*Below:* Sam Esler, one of a great family of South Island horsemen, showing good style on a dirty one. *G. Roberts*

*Above:* The New Zealand jumping team captained by Florence West at Auckland in 1974. *From left:* Florence West and Dusky Bay, Alexis McHardy and Tapioca, Pam Bailey and Parnell, Mary Watkinson and Beau de L'aire. *Pam Howlett*

*Below:* Jill Newton, aged seven, with a Timor foal belonging to Mrs Rosemary Page. *P. Newton*

on foot. In the main the back country consists of high rolling tops about the 5000 ft mark, and even on horseback a man's vision is so restricted that a tremendous amount of riding is involved. With the front facings rising steeply onto the tops, the horses have a big climb each morning. All told it's no picnic. The two huts from which the mustering of the Nevis is done both have "road" access and feed is carted out for the horses – but even so they have a pretty lean time.

I took Bill down to Nokomai with me and he earned quite a name for himself. However when I came back to Canterbury he was getting on in years so I sold him to a well-known Southland farmer, Guy Chewings of Mossburn. Bill suited him down to the ground and I knew that he'd have a good home for the rest of his days.

It was while I was at Nokomai that one of the musterers came mighty close to losing his life – not to mention that of his horse – through his own foolhardiness. At the time Jock McCauley, a well-known local musterer, was camped out at the Nevis boundary keeping. When the mustering season was finished at the end of May this young fellow wanted to go out and join him. This involved a 20-mile ride – up the Nokomai creek, over the Nevis Saddle and down the Nevis River to the Nevis hut. There was quite a bit of snow in the saddle at the time, but John, the young fellow, was quite confident that this would present no problems. Waiting until conditions were suitable, he took off.

However there was more snow than he had thought. After topping the saddle his hack – a very nice mare that he thought the world of – was soon in difficulties. Loath to leave her, John battled away for several hours with little progress and finally, when she got hopelessly bogged in one of the many mining excavations that dotted that locality, he had no option but to abandon her and continue on foot. As it happened there was a plaster of snow in the Nevis and John had a tremendous struggle. He was actually the best part of 24 hours doing that 20 miles, and at times he thought he'd never make it. When he finally reached the hut he was completely exhausted and suffering quite badly from frostbite. Thanks to Jock McCauley's care he suffered no lasting ill effects, but it was a close call.

Fortunately his mare came to no harm either. Jock went looking for her a couple of days later to find that she had scrambled out of that hole and was waiting nearby completely unharmed.

Jock McCauley, who was then one of our top dog trial men, had two nice horses while I was with him. One was his hack, a classy sort of a mare, the other a very nice chestnut cob called Darby. A sturdy little horse, he was both a good hack and a good pack-horse. Jock eventually sold him to the station and he was a real acquisition. He hardly matched the pack-team, which consisted of big, fairly heavy horses, but he was an ideal packies' hack – level-headed, sensible and quiet – and of course could also be packed if necessary.

It was while at Nokomai that I first met that well-known character "The Bearded Horseman". Commonly known as Johnny Walker, he travelled the roads on horseback from one end of the island to the other and was well known throughout the back country. He usually had a dog following him and he'd stop off every here and there, do a few days work and then move on again. My first meeting with him – at Nokomai – was rather amusing. One morning I heard the tinkle of hammer and anvil across at the blacksmith's shop. Knowing that none of the station hands were about I went across to investigate. To my surprise I found a black bearded man, a complete stranger, busy shoeing his horse. Apparently he had pulled in the previous evening, helped himself to a bed in one of the huts, lined up for breakfast at the cookshop the next morning, and without so much as bye your leave, headed down to the blacksmith's shop to shoe his horse.

"Johnny Walker" became well known in this district – Rangiora-Whiterock – in later years. He was always well mounted, and by all accounts he was a pretty good hand with a horse.

Another amusing incident concerns a couple of the Nokomai musterers. The gang was to go out to the Nevis this day to start the autumn muster and from the stable they had to pass my house. I saw them ride past, in ones and two, until finally these two appeared leading their horses. It seems that they had been celebrating a birthday or something – and making a pretty good job of it – for they were having a fair bit of difficulty in getting mounted. One of them, a little less the worse for wear, would give his mate a helping hoist as he endeavoured to climb aboard, but this only resulted in the would-be rider toppling clean over the top of his horse and landing in a heap on the far side. This happened several times and they were getting nowhere fast. However the shock of his repeated falls seemed to sober this bloke up and as his mate seemed to have taken a turn for the worse he himself was in need of assistance.

In the end it was hard to say who was helping who, but they were tryers, those blokes, and they finally got mounted. Away they went as happy as Larry.

I enjoyed my sojourn in Southland – they're great people down there – but our stay was only a brief two years because I was offered an attractive salary to go back up to my old home, Mt White. Mt White, which totals 120,000 acres, is walking country, but it was still one of the few Canterbury stations that had not gone in for the landrover and it still had a pack-team in regular use. They were no longer breeding draughts, but they still kept a light stallion and bred their own hacks. However this horse had little to recommend it and the line-up of station hacks was not very impressive.

As I did not arrive there until some time after the season had started each of the musterers had been allotted his hack and I had to take what was left. It was a slug of a mare called Strawberry. Old "Strawb", as she was called, did have her good points, although speed and activity weren't among them. But at least she always got there in time. There's always a place for a horse like her on a station – anyone at all could take her with safety and the very poorest of horsemen could never spoil her.

However among the station string of hacks was a very nice bay cob named Nobby. He was getting on in years but was still an active, free moving horse and when the musterers finished up at the end of the season I made no bones about claiming him as my hack. He had that most desirable of gaits, the amble, and for all his age was lovely to sit on. I don't know where he came from but as a youngster he must have been a beauty. I rated that horse as one of the best I've ridden.

It was while at Mt White that I saw a feat of horsemanship that would take a lot of beating. At the time the string of young horses included four three-year-olds that were about ready to break in and the upshot was that the musterers undertook to take one each. The trouble was that there were five musterers and only four horses; someone had to miss out and as was to be expected this was a young fellow who had only recently joined the gang. However he was made of good stuff, this bloke, and he decided he wasn't going to miss out. He'd been at Lake Sumner Station before coming to Mt White and knowing that they had a string of young horses he calmly announced that he'd slip over there and get one. He was a quiet, retiring sort of a chap and no-one took him seriously – it's all of 40 miles to Lake Sumner, with a high range to cross. Anyway one morning when there was nothing much doing he saddled up a hack and stuffing a bit of tucker in his saddle bag he took off.

The first day he rode up the Esk as far as Anderson's camp and crossing over the Dampier Range – the boundary between Mt White and Eskhead – he camped at Deep Creek hut on the latter station; it was a ride of about 25 miles. The next morning he completed the ride to Lake Sumner and in the afternoon ran in the young'ens and selected and caught the one he wanted. He spent the third day handling, shoeing and teaching it to lead and on the afternoon of the fourth day he headed back to Deep Creek, snigging the young'en behind him. He landed back at the Mt White homestead on the afternoon of the fifth day after towing an unbroken youngster up hill and down a full 40 miles. I'd say that was quite an achievement.

I might add that the horse turned out to be a real plaster. It was, I believe, finally shot for dog tucker.

I occasionally called in to see Oscar Watson at Rosewood while I was at Mt White. On one occasion I discovered that he'd recently had Red Chase put down. I was sorry to hear this, as he was still an entire and I'd have jumped at the chance to acquire him for the station. Although only a little over 15 hands he was a beautifully coupled, strongly built horse, with guts to burn. He would have been ideal for breeding station hacks. There is a lot to be said for the trotting horse in back country; they "do" better under harsh conditions, they are level headed and they are more suited to the various assortment of riders that may have to handle them. Best of all it is from the trotter that we get the ambler. This is the ideal gait for a shepherd. The horse is easy to sit on and although the pace is considerably faster than a walk it is not too fast for dogs to follow.

Even though 80 percent of the country is mustered on foot there is still a lot of riding on Mt White. From the top end of the station, at the Waimakariri railway bridge at Cora Lyn, to the far boundary at the head of the Esk River involves a ride of 50 miles, and although the station is carrying only 400 head of cattle they are spread over the length of the property. All told the horse plays an important part in the running of the station. As I remarked earlier, their horses have slipped quite a bit since the days of the old "Mt Whiters". With his spirit and constitution a horse like Red Chase could have been just the answer.

# 14. The sporting life

WITH my children's education becoming a problem I finally migrated back down to the "mud flats". Once again I thought that this would virtually spell the end of my active association with the horses. With the advent of the tractor – that unsociable monstrosity – team-horses were now almost a thing of the past; the landrover – and even the helicopter – was slowly but surely taking the place of the shepherd's hack. It looked as if the horse had about had its day.

But such was not the case. This was simply the start of a new era for it was about this time – in the 60s – that there was a tremendous upsurge in pleasure riding. Pony clubs boomed dramatically, show jumping became one of our major sports and in Canterbury in particular polo was once again coming into its own. Rodeos, horse treks and endurance riding were all to play a notable part in this trend. Today the horse probably has more followers and admirers than ever before. It is good to see.

For my part I eventually took on a little grazing run – about the size of the Mt White horse paddock – at Whiterock, some 15 miles from Rangiora. It involved only two easy "days" mustering but I occasionally gave my neighbours a hand. And of course I had to have a hack. I was lucky enough to be able to buy one on the place when I took over; a big rangy gelding standing about 17 hands, it was said to be by a well-known trotting sire, Steel King. For all its faults it was quite a remarkable horse. Although getting on in years, and as mad as a maggot in company, it was a tremendous walker and when it came to climbing I've never seen anything to compare with him. He just went at a hill like a mad thing and left everything else miles behind. He'd have killed himself if I'd let him go.

I had old Wisdom, as he was called, for ten years. Although he must then have been well past 20 he was still as game as ever. He served me well and I've a pretty soft spot for that old fellow. There were two other hacks in that locality that I rate among the good ones I've known. One was a great big heavy hack – a pack-horse type – that was owned by Harry Gudex, a neighbour of mine. Harry was a man of 15 stone or more and old Master, as his horse was called, carried him up hill and down dale without turning a hair. A fine stamp of a horse, level headed and sensible, he was a grand hack. The other was a very attractive cream cob owned by Bevan Rossiter, a local boy who used to give me a hand with my mustering. A great stamp of a hill horse, he would go anywhere a horse could possibly go. Even in the worst of going Bevan seldom thought it necessary to get off and lead him. That little horse was just about as good on the hill as any I've seen.

Another memory of the Whiterock horses concerns a mare belonging to Lester Lang, another neighbour of mine. Although as pretty as a picture this mare was as cranky as they come, and even at the best of times Lester had difficulty in controlling her. On this particular occasion Lester had given me a hand to muster and we were quietly dawdling along behind the mob when his mare trod fair on top of a wasps' nest. There was a howl of anguish from Lester as one of the wasps presented him with its visiting card. The mare, with wasps attacking her on all sides, took off like a bat out of hell – how Lester stayed with her I wouldn't know. We were crossing a dry creek bed at the time and for a few moments I couldn't see man or horse for flying shingle. That mare really out-did herself that day.

Fortunately I was a chain or two away at the time and I thought it quite funny. I gathered that Lester didn't, but I'm afraid I can hardly record his comments – much as I'd like to.

I had another experience while at Whiterock which was a striking illustration of how much lack of confidence in a rider can upset a horse. Feeling that Wisdom was getting past the job I purchased another hack; although it was just a quiet old prad when I got it that horse eventually went completely haywire with me. Always inclined to be a bit nervy with a horse, I found that as I grew older this nervousness increased until finally what little confidence I might have had disappeared almost completely. Anyway this horse became virtually terrified of me and at times I had difficulty in saddling him. Strangely enough he never attempted to buck, but he was such a bundle of nerves whenever I rode him that he was practically useless. I finally sold him, and although his new owner was comparatively inexperienced with horses he settled down with him and was as good as gold.

As it happened the Whiterock-Loburn district was then a real stronghold of equine enthusiasts, the best known at that time being Mrs Diana Bain; her husband, Russell, was headmaster at the North Loburn school. At her best Diana Bain was probably the leading South Island lady rider. Over the years she won practically every competition open to her.

When I was first at Whiterock her star horse was a little chestnut gelding named Harvest Moon, who at only 14.3 was just out of pony class. Even with this handicap he twice won the A.1. Cup, which includes dressage, cross-country and show jumping and is regarded as the premier New Zealand one-day event for riders aged from 17 to 21. Her next horse was Hunter's Moon, a very attractive grey gelding. A beautifully mannered level headed horse, he excelled at show jumping and won many major events in this field. Her present horse, a chestnut mare named Juanita Banana, is probably the best she has had. With her best years still ahead of her, this mare has already established herself as a show jumper second to none. It was with her that Diana won the 1977 New Zealand Lady Rider of the Year award. This was a fitting reward for her years of service to this most popular of sports.

Another local rider to meet with considerable success is Dennis Youngman, also of Loburn. Although he has been one of our leading show riders for a number of years it was with the chestnut mare Fiano that he really reached the top flight. In 1961, when Dennis was only 17 years of age, they were included in the 13 that were selected to go into training as possible candidates for the New Zealand team to go to the Tokyo Olympics. Unfortunately Fiano broke a bone in her foot while on that training course and this put her out of the reckoning. However she eventually recovered and jumped with marked success in New Zealand for several years after that. Altogether Dennis represented Canterbury on eight occasions – three of them with Fiano. His present horse is a beautiful, big six-year-old grey mare named Annie Lawry and she also is showing a lot of promise. Her most recent success was at Methven when she won the Futurity Stakes.

Dennis's wife, Denise, is also a keen horsewoman. For some years she figured among our leading lady riders and with Fleetwing had the distinction of representing Canterbury on two occasions. Her most notable contribution to the game is that for the last 17 years she has been one of the three course planners for all show jumping events throughout Canterbury.

Reg Youngman, Dennis's father, is also a prominent figure in show jumping circles. As the mounted marshall for all the Christchurch and Rangiora jumping fixtures as well as all horse society events in Canterbury he is well known to all jumping enthusiasts.

With the tremendous upsurge in pony club activities taking place about this time, Whiterock and the surrounding districts became very popular for riding trips. No-one was more active in fostering the movement than Mrs Rosemary Page of Mt Lawry. One of the most favoured trips was across to Glentui, over into Lees Valley via the Blowhard track, and out to Whiterock via The Brothers and Okuku Pass. This was a five or six-day trip and with anything up to 30 riders the organisation was a big undertaking. Rosemary organised and led those parties for about 15 years.

She and her husband Brian also breed Timor ponies – just about the smallest of all breeds and fascinating little animals. Their stallion Peanut is by Mrs Daisy Andrew's Wee Geordie, who was sired by a Timor pony that once toured this country with Wirth's Circus. Peanut, who stands only ten and a half hands, is a beautiful little horse and has left some lovely stock.

Brian, a keen supporter of polo, was largely instrumental in the founding of the Ashley Polo Club, which at one time fielded a team that compared with any in the island. One of their players, Derek Wells of Oxford, represented the South Island at the Christchurch Show on three occasions. Des Lundy – whose death at only 32 years of age was one of the tragedies of this district – was also a top player.

But probably none were more widely known as horsemen than the Esler family. Jack Esler, another who died before his time, hailed from over Murchison way, although he spent many years mustering in the Canterbury back country. He purchased Okuku Pass in 1951 – the year of the wool boom – and was soon to establish himself as one of the stalwarts of the rodeo game. The whole family – Jack himself, his two daughters, Nancy and Florence, and his son Ian – all made a name for themselves in that most testing of "pastimes". Jack, who was then past rough riding, was prominent in the organisation side of the game but "Sam", as Ian was known, became a top competitor both at riding and bull dogging. He also competed in both Australia and North America. Nancy and Florence could also sit a rough horse with the best of them and they were also prominent in the show ring.

Florence – now Mrs Garnett West – was later to establish herself, both at show jumping and over country, as a rider second to none. She represented New Zealand in Australia in 1973 and the following year was appointed captain of the New Zealand team which competed against the Aussies at Auckland. On both occasions she came through with flying colours.

Her present horse, Dusky Bay, needs no introduction to anyone interested in jumping. A bold, upstanding brown gelding he was purchased by Florence unbroken for $47 at a St James horse sale. Although he took a lot of handling he was to prove himself one of the most versatile jumping horses in the country. In 1973 he won that most testing of all events, the Three Day Trial, and the following year won the South Island points trophy. This is awarded to the horse and rider scoring the most points at all horse trials during that season. Suffice it to say that although now only 11 years of age he has already won just on $3000 and is still going like a train. Surely he can justly be rated one of the greatest event horses this country has produced.

Jack Esler is a tough, hardy back country man who made his mark as a horseman when he first came to Whiterock. With two assistants – Rod Woodham, an old Murchison identity, and a lad named Mervyn Iles – he drove a mob of 20-odd horses, mostly unbroken, from his farm at Tutaki through to Okuku Pass – a distance of about 150 miles. For practically the whole distance they travelled through open country. This must surely rank as one of the great droving feats of modern times.

From Tutaki, which lies halfway between Murchison and Lake Rotoroa in southern Nelson, they cut through to the Oxnam's Mataki Station, followed down the Matakitaki River to Glenroy and over the Maruia Saddle. From there they followed up the Maruia River, past Springs Junction and over the Lewis Pass. At the junction of the Boyle and the Hope they turned up the latter river, crossed the Kiwi Saddle in heavy bush and carried on through to Lake Sumner. They then followed down the Hurunui as far as the Seaward. This led them out over the Virginia country into the Waitohi. From there they cut through McDonald Downs to Whiterock. At this time – 1951 – droving was practically a thing of the past and all stock was carted by transport. Not many men would even have tackled that drive.

Another Esler achievement that is even more noteworthy concerns Samuel Esler, Jack's father, when the family first went to Tutaki away back in 1910. At the time Sam Esler was farming at Albury in South Canterbury. Hearing that land was being opened up for settlement in Westland and southern Nelson he decided to investigate the possibilities. In August of that year he and his wife left Albury in a horse and gig. Travelling via Arthur's Pass they eventually got as far as Murchison. It was there, in the Tutaki Valley, that Esler found the place he wanted; making a deal on the spot they then returned to Albury – a journey all told of about 600 miles.

Esler had already sold his Albury property, and acquiring a covered waggon and two horses he loaded up his family and their personal belongings and once again set out for Tutaki. This time they headed up through Hanmer Springs, over Jollies Pass and up the Acheron to Tarndale, hoping that they would be able to follow down the Wairau to Top House. This proved to be quite impossible and they were forced to go down the Awatere and out to Blenheim, although this involved another 150 miles. From there they headed up the Wairau to Top House and down the Buller to their destination. This journey, estimated to be 500 miles took them exactly a month – a tremendous undertaking under the conditions of those times.

Jack Esler was then seven years of age; his brother Andrew was nine and his sister Nancy, now Mrs M. A. Lammas of Rotheram, was five. The cavalcade included a pony, which the two boys took turns at riding. I'd say the Eslers have got what it takes. I might add that the Crown recently took over Okuku Pass as a development project and appointed a local man, Lyn Morris, as manager. Lyn is one of our top rodeo riders and he also is giving the game valuable support. Garnett West himself was once a leading rider. Another ex-Whiterock man to reach the top flight is Ian (Taffy) Henry, now of Christchurch. Taffy, who had his grounding with the Eslers at Okuku Pass, is the present national bareback champion – according to Florence West he was a born natural rider right from the start.

Murray Curle of Oxford is yet another local man to make his mark in rodeo circles. A prominent rider in his younger days, he later concentrated on roping events. His horse, Socks – a smart little grey gelding – is well known to rodeo enthusiasts throughout the island. In recent years Murray has set himself up as a "stock contractor"; with 30-odd bucking horses, one third of which are his own, he provides most of the horses used in buck jumping events in Canterbury. He keeps them in great fettle and they are considered to be as good a string as any in the country. I might explain that a "good" horse is one that consistently puts up a good bucking performance; indeed at some rodeos a ribbon is awarded to the best bucking horse at the meeting. Such horses are keenly sought after by the contestants because unless a rider draws a "good" horse he hasn't the opportunity to demonstrate his ability and win the event.

Murray's latest move is the acquisition of a dozen young Brahman cross bulls which he plans to make available for bull riding events. The Brahmans are reckoned the most difficult of any to ride; these bulls will test our riders to the full and should add still further to the attractions of the game.

Murray's chief assistant is Ransom Bruce – one time of Lees Valley – who in his day was one of our outstanding riders. Although at the time rodeo championships had no official status he twice competed with marked success at the world series held in Australia back in the early 50s and was recognised as one of the great riders of his day. He is also an expert at adjusting the bucking strap – or flank strap – which is one of the most important of all jobs at a rodeo. It is the adjustment of this strap which more than anything determines the extent to which a horse will perform. This of course makes or mars a rodeo.

With the sport now under the jurisdiction of a world body, championship events have official recognition and rodeos are increasing in popularity every year. Including island and national finals there are well over 30 affiliated fixtures held in this country every year. With its full share of spills and thrills this is a *man's* game. Its appeal to the public is such that at some meetings the number of spectators runs into many thousands. The organising and running of these fixtures is a big undertaking.

Although my interest in racing had gradually waned since my Rakaia days I still attended an odd trotting meeting and it was while at Whiterock that I had the pleasure of seeing the immortal Cardigan Bay win the New Zealand Trotting Cup at Addington. This was in 1963, when at seven years of age this mighty horse was probably at his best. During that week he won four major events, including the New Zealand Free for All. His mile ratings – over distances ranging from one and a quarter to two miles – varied only from 2.02 to 2.05 $^{3}/_{5}$. Those four races earned him $29,200.

On his record Cardigan Bay must take pride of place as the greatest pacing horse this country has produced. All told he won 80 races in his ten years of racing and his stake earnings amounted to the incredible sum of $1,000,837. He was the first horse ever to pass the million-dollar mark and needless to add he was then rated as the greatest pacing horse of all time. Of his total earnings $73,986 was amassed on

New Zealand tracks, \$49,880 in Australia and \$843,141 in the United States. His best times were 1.56½ for the mile – against time – which was a world record for a gelding – and 4.10⅗ for the two miles.

At seven years of age he was sold to United States interests as his owner Mrs Audrey Dean felt that he would not have the scope in this country to fully prove his worth. However it was stipulated that when his racing days were over – or when he reached 13 years of age – he was to be returned to New Zealand. There were some who criticised Mrs Dean for selling him but had she not done so this mighty horse would never have had the opportunity to become The Million Dollar Horse.

Over the years we have had several horses that showed just as much potential as Cardigan Bay, but they were never given the opportunity. Highland Fling, for one, had a mighty record, but he had the misfortune to suffer an injury that put him out of racing when only six years of age. His winnings – on New Zealand tracks – were then considerably in excess of Cardy's at that age, and his best times were almost exactly similar. Harold Logan and Johnny Globe were others who, given the chance, could have vied with any. World records have been established by numerous New Zealand horses and although nothing can detract from the peerless achievements of Cardigan Bay it must be conceded that thanks to Mrs Dean he was given the opportunity to prove his greatness. The others weren't.

It was about this time – the mid 60s – that Jim Stevenson decided to hold his bi-annual sale of horses at the St James homestead. The sale is held in February and at that time of the year this is a lovely spot; the homestead is situated right alongside the Clarence River and with towering tops on all sides it has a real back country atmosphere. The crowd, which includes prominent horsemen from all corners of the province, makes this fixture a real picnic outing and as it is only eight miles from Hanmer – via Jacks Pass – it is a popular event.

One of the real personalities of the gathering is Miss Rosemary Grimsey of Oxford. For a number of years Rosemary – who is also well known in the show ring – has regularly given a hand to muster and sort out the sale draft. She too can sit a horse with the best of them, and with her sunny disposition she is one of the most popular figures in the game.

Rosemary's father, Mr Ted Grimsey, who has since died, was also a regular attendant at this sale. Ted, who was prominent in Clydesdale circles in his younger days, actually drove a team for over 50 years. When he returned from the Boer War he took over a small farm at Ashley – in the Rangiora district – and to supplement his income he did contract work with the horses. He later took over his present farm at Oxford and actually drove a team until he finally gave up the horses some time after World War II. His association with the Clydesdale horses must surely be unequalled.

For some years Jim Stevenson has been using only thoroughbred sires at St James and the sale horses are now light, clean legged types. They are keenly sought after by jumping enthusiasts. With the toughness which is a heritage of the harsh back country conditions under which they are reared the St James horses can more than hold their own under the severest of tests. Many of them have become top horses in both the show ring and horse society events, the most notable, of course, being Florence West's Dusky Bay. Two others of note are Alan Binnie's Morgan and Malcolm McCall's Pegasus – both of which were sold to Japanese interests while still in their prime.

A mare that met with marked success in the show ring is Miss Judy Stevenson's Troy; a bright chestnut with a white blaze and four stockings, she is a lovely little mare. For four years on end she was the champion cob at the Christchurch Show, and she also has done much to enhance the reputation of the St James horses.

St James is the only back country station in this island that is still breeding and selling horses on a worthwhile scale. Anything up to 70 are offered at each sale, and horsemen and women throughout Canterbury – and from further afield – will be hoping that they will always be available.

# Epilogue

WHEN I finally retired into Rangiora that was truly the end of my active association with the horses. However, when I was writing of the New Zealand stations, my travels took me into many new corners. I was to see wild horses on Ohinewairua Station, north of Taihape and I also saw something of the adjoining Ngamatea country. Ngamatea Station, although officially "only" 120,000 acres, bounds onto a vast track of "no man's land" at the head of the Rangitikei and Ngaruroro Rivers and the stock range over probably double that area. The handling of this mighty holding involves more riding than on any other single property in the country and their string of horses totals a hundred or more.

I was also to visit the Sisam's Awakatoro Station in the Bay of Plenty. This fabulous property still runs four teams of working draughts and they also have an Arab stud. They have some lovely horses; the two-year-olds – a dozen or so – running loose on a block of beautiful rolling country, interspersed with native bush, were an unforgettable sight. In the far south I was to meet Dick Restall and his famous "Donkey". This little horse, a halfbred Arab, made history by twice winning that toughest of all tests, the 50-mile endurance event held at Burwood Station in Southland. A mighty little horse.

All told I have had the good fortune to know, and see, many wonderful horses. I feel that in recording something of my experiences I have been able to pay them the tribute they so richly deserve.

# Index

# OTHER REED BOOKS

Peter Newton has written a series of four regional surveys recording the great New Zealand farm runs and the men and women who farm them.

The books describe the country's most outstanding farm properties, recording their histories to the present day, their current stocking and management policies, and drawing on the memories of surviving oldtimers. The stories are told with a high-countryman's understanding of and affection for country matters. The result is both fascinating and informative.

**Big Country of the North Island** – 45 prominent stations in the eastern North Island, from White Rock in the southern Wairarapa to Waipoa north of Gisborne.

**In the Wake of the Axe** – the big country of the west and north of the North Island, from Orongorongo in the Wellington district to Te Paki at the very tip of the North Island.

**Big Country of the South Island** – the South Island north of the Rangitata: stations in Nelson, Marlborough and North and Mid-Canterbury.

**Sixty Thousand on the Hoof** – the big country south of the Rangitata, through Mid and South Canterbury, Westland, Otago and Southland, areas where the majority of the truly great runs are maintained.